TRANSPORT PHENOMENA
IN BIOMEDICAL
ENGINEERING
PRINCIPLES AND PRACTICES

TRANSPORT PHENOMENA IN BIOMEDICAL ENGINEERING
PRINCIPLES AND PRACTICES

Edited by

Robert A. Peattie
Robert J. Fisher
Joseph D. Bronzino
Donald R. Peterson

CRC Press
Taylor & Francis Group
Boca Raton London New York

CRC Press is an imprint of the
Taylor & Francis Group, an **informa** business

CRC Press
Taylor & Francis Group
6000 Broken Sound Parkway NW, Suite 300
Boca Raton, FL 33487-2742

© 2013 by Taylor & Francis Group, LLC
CRC Press is an imprint of Taylor & Francis Group, an Informa business

No claim to original U.S. Government works

Printed in the United States of America on acid-free paper
Version Date: 20120823

International Standard Book Number: 978-1-4398-7462-2 (Hardback)

Library of Congress Cataloging-in-Publication Data

Transport phenomena in biomedical engineering : principles and practices / edited by Robert A. Peattie ...
 [et al.].
       p. ; cm.
 Includes bibliographical references and index.
 ISBN 978-1-4398-7462-2 (hardcover : alk. paper)
 I. Peattie, Robert A.
 [DNLM: 1. Biological Transport--physiology. 2. Body Fluids--metabolism. 3. Cell
 Membrane--metabolism. 4. Drug Delivery Systems. 5. Hydrodynamics. QU 120]

 571.6'4--dc23   2012033796

Visit the Taylor & Francis Web site at
http://www.taylorandfrancis.com

and the CRC Press Web site at
http://www.crcpress.com

Contents

Introductory Comments vii
Robert J. Fisher

Editors xiii

Contributors xv

1   Biomimetic Systems: Concepts, Design, and Emulation 1-1
     Robert J. Fisher

2   Transport/Reaction Processes in Biology and Medicine 2-1
     E. N. Lightfoot

3   Microvascular Heat Transfer 3-1
     James W. Baish

4   Fluid Dynamics for Bio Systems: Fundamentals and Model Analysis 4-1
     Robert A. Peattie and Robert J. Fisher

5   Animal Surrogate Systems 5-1
     Michael L. Shuler, Sarina G. Harris, Xinran Li, and Mandy B. Esch

6   Arterial Wall Mass Transport: The Possible Role of Blood Phase
    Resistance in the Localization of Arterial Disease 6-1
     John M. Tarbell and Yuchen Qiu

7   Transport Phenomena and the Microenvironment 7-1
     Robert J. Fisher and Robert A. Peattie

8   Transport and Drug Delivery through the Blood–Brain Barrier
    and Cerebrospinal Fluid 8-1
     Bingmei M. Fu

9   Interstitial Transport in the Brain: Principles for Local Drug Delivery 9-1
     W. Mark Saltzman

10 Surfactant Transport and Fluid–Structure Interactions during
    Pulmonary Airway Reopening 10-1
     David Martin, Anne-Marie Jacob, and Donald P. Gaver III

Index Index-1

Introductory Comments

Robert J. Fisher

The intention of this book is to couple the concepts of transport phenomena with chemical reaction kinetics and thermodynamics to introduce the field of reaction engineering. This is essential information needed to design and control engineering devices, particularly flow reactor systems. Through extension of these concepts, combined with materials design, to mimic biological systems in form and function, the field of biomimicry has evolved. The development of biomimetic systems is a rapidly emerging technology with expanding applications. Specialized journals devoted to exploring both the analysis of existing biological materials and processes and the design and production of synthetic analogs that mimic biological properties are now emerging. These journals blend a biological approach with a materials/engineering science viewpoint covering topics that include analysis of the design criteria used by organisms in the selection of specific biosynthetic materials and structures; analysis of the optimization criteria used in natural systems; development of systems modeled on biological analogs; and applications of "intelligent" or "smart" materials in areas such as biosensors, robotics, and aerospace. This biomimicry theme is prevalent throughout all the chapters in this book, including one specifically devoted to these concepts with explicit examples of applications to reacting and transport processes.

The field of transport phenomena traditionally encompasses the subjects of momentum transport (viscous flow), energy transport (heat conduction, convection, and radiation), and mass transport (diffusion). In this text, the media in which the transport occurs is regarded as continua; however, some molecular explanations are discussed. The continuum approach is of more immediate interest to engineers, but both approaches are required to thoroughly master the subject. The current emphasis in engineering education is on understanding basic physical principles versus "blind" use of empiricism. Consequently, it is imperative that the reader seek further edification in classical transport phenomena texts: general (Deen, 1996; Bird et al., 2002), with chemically reactive systems (Rosner, 1986), and more specifically, with a biologically oriented approach (Lightfoot, 1974; Fournier, 1999). The laws (conservation principles etc.) governing such transport will be seen to influence (1) the local rates at which reactants encounter one another; (2) the temperature field within body regions (or compartments in pharmacokinetics modeling); (3) the volume (or area) needed to accomplish the desired turnover or transport rates; and (4) the amount and fate of species involved in mass transport and metabolic rates. For biomedical systems such as dialysis in which only physical changes occur, the same general principles, usually with more simplifications, are used to design and analyze these devices. The transport laws governing nonreactive systems can often be used to make rational predictions of the behavior of "analogous" reacting systems. The importance of relative time scales will be discussed throughout this book by all authors. It is particularly important to establish orders of magnitude and to make realistic limiting calculations. Dimensional analysis and pharmacokinetic modeling techniques are especially attractive for these purposes; in fact, they may permit unifying the whole of biological transport (see other chapters of this text and Enderle et al., 1999; Fisher, 1999). As the reader progresses, the importance of transport phenomena in applied biology becomes steadily more apparent. "In all living organisms, but most especially the higher animals, diffusional and flow limitations are of critical importance; moreover, we live in a very delicate state of balance with respect to these two processes" (Lightfoot, 1974).

Each chapter in this book is largely self-contained. Similar concepts are brought forth and reinforced through applications and discussions. However, to further enhance the benefits obtained by the reader, it is prudent to first discuss some elementary concepts; the most relevant being control volume selection and flow reactors.

When applying the conservation laws to fluid matter treated as a continuum, the question arises as to the amount of matter to be considered. Typically, this decision is based on convenience and/or level of detail required. There is no single choice. Many possibilities exist that can lead to the same useful predictions. The conservation laws of continuum dynamics can be applied to the fluid contained in a volume of arbitrary size, shape, and state of motion. The volume selected is termed as "control volume." The simplest is one where every point on its surface moves with the local fluid velocity. It is called a "material" control volume since, in the absence of diffusion across its interface, it retains the material originally present within its control surface. Although conceptually simple, they are not readily used since they move through space, change their volume, and deform. An analysis of the motion of material control volumes is usually termed "Legrangian" and time derivatives are termed "material" or "substantial" derivatives.

Another simple class of control volumes is defined by surfaces fixed in physical space, through which the fluid flows. These "fixed" control volumes are termed as "Eularian" and may be either macroscopic or differential in any or all directions. The fluid contained within an Eularian control volume is said to be, in thermodynamic terms, an "open" (flow) system.

The most general type of control volume is defined by surfaces that move "arbitrarily," that is, not related to the local fluid velocity. Such control volumes are used to analyze the behavior of nonmaterial "waves" in fluids, as well as moving phase boundaries in the presence of mass transfer across the interface (Crank, 1956; Fisher, 1989).

Characterization of the mass transfer processes in bioreactors, as used in cell or tissue culture systems, is essential when designing and evaluating their performance. Flow reactor systems bring together various reactants in a continuous fashion while simultaneously withdrawing products and excess reactants. These reactors generally provide optimum productivity and performance (Levenspiel, 1989; Freshney, 2000; Shuler and Kargi, 2001). They are classified as either tank- or tube-type reactors. Each represents extremes in the behavior of the gross fluid motion. Tank-type reactors are characterized by instant and complete mixing of the contents and are therefore termed as perfectly mixed, or backmixed, reactors. Tube-type reactors are characterized by the lack of mixing in the flow direction and are termed plug flow, or tubular, reactors. The performance of actual reactors, though not fully represented by these idealized flow patterns, may match them so closely that they can be modeled as such with negligible error. Others can be modeled as combinations of tank- and tube-type over various regions.

An idealized backmixed reactor is analyzed as follows. Consider a stirred vessel containing a known fluid volume into which multiple streams may be flowing that contain reactants, enzymes (biocatalysts), nutrients, reaction medium, and so on. When all these components are brought together in the vessel, under properly controlled conditions such as temperature, pressure, and concentrations of each component, the desired reactions occur. The vessel is well mixed to promote good contacting of all components and hence an efficient reaction scheme can be maintained. The well-mixed state is achieved when samples withdrawn from different locations (including the exit) at the same instance in time are indistinguishable. The system is termed "lumped" versus "distributed," as in a plug flow system where location matters. The response characteristics of a backmixed reactor are significantly different from those of a plug flow reactor. How reaction time variations affect performance and how quickly each system responds to upsets in the process conditions are key factors. The backmixed system is far more sluggish than the plug flow system. To evaluate the role of reaction time variations, the concept of residence time and how it is determined must be discussed. A brief analysis of a batch reactor will be useful in understanding the basic principles involved.

In batch systems, there is no flow in or out. The feed (initial charge) is placed in the reactor at the start of the process and the products are withdrawn all at once at some later time. Spatial uniformity of composition is assumed since vigorous mixing is typically applied. If accomplished (ideally), then the time for reaction is readily determined. This is significant since the conversion of reactants to products is a function of time and can be obtained from knowledge of the reaction rate and its dependence upon composition and process variables. Since these other factors are typically controlled at a constant value, the time for reaction is the key parameter in the reactor design process. In a batch reactor, the concentration of reactants is time dependent, and therefore, the rate of reaction as well. Conversion is now related to reaction time through the use of calculus. For this system, residence time, equal to the processing time, is the time for reaction.

In a backmixed reactor, the conversion of reactants is controlled by the average length of time fluid elements remain in the reactor (their residence time). The ratio of the volume of fluid in the tank to the volumetric flow rate of the exit stream determines the residence time. Recall that in an ideal system of this type, operating at steady state, concentrations in the vessel are uniform and equal to those in the exit stream and thus the reaction rate is maintained at a constant value. Conversion of individual reactants and yield of specific products can be determined simply by multiplying the appropriate rate of interest by the residence time. With imperfect mixing, fluid elements have a distribution of residence times and the performance of the reactor is clearly altered.

Analysis of the plug flow reactor system is based on the premise that there is no mixing in the flow direction and thus no interaction between neighboring fluid elements as they traverse the length of the reactor. This idealization permits use of the results obtained from the analysis of the batch reactor system. Each fluid element in the tubular reactor functions as a small batch reactor, undisturbed by its neighboring elements, and its reaction time is well defined as the ratio of tube length to the volume-averaged fluid velocity. Concentration varies along the length of the reactor (also, rate) and can be simply related to reaction time. The mathematical analysis and prediction of performance is similar to that for a batch reactor. This analysis shows that the plug flow configuration yields higher conversions than the backmixed vessel, given equal residence times and the same processing conditions. However, the plug flow reactor responds more quickly to system upsets and is more difficult to control.

Most actual reactors deviate from these idealized systems primarily because of nonuniform velocity profiles, channeling and bypassing of fluids, and the presence of stagnant regions caused by reactor shape and internal components such as baffles, heat transfer coils, and measurement probes. Disruptions to the flow path are common when dealing with heterogeneous systems, particularly when solids are present. To model these actual reactors, various regions are compartmentalized and represented as combinations of plug flow and backmixed elements. For illustration, a brief discussion of recycle, packed bed, and fluidized bed reactor systems are follows.

Recycle reactors are basically plug flow reactors with a portion of the exit stream recycled to the inlet, which provides multiple passes through the reactor to increase conversion. It is particularly useful in biocatalytic reactor designs, in which the use of a packed bed is desired because of physical problems associated with contacting and subsequent separation of the phases. The recycle reactor provides an excellent means to obtain backmixed behavior from a physically configured tubular reactor. Multiple reactors of various types, combined in series and/or parallel arrangements, can improve performance and meet other physical requirements. Contact patterns using multiple entries into a single reactor can emulate these situations. A system demonstrating this characteristic is a plug flow reactor with uniform side entry. If these entry points are distributed along the length with no front entry, the system will perform as a single backmixed system. The significance is that backmixed behavior is obtained without continual stirring or the use of recycle pumps. Furthermore, if these side entry points are limited to only a portion of the reactor length, then the system functions as backmixed reactors in series with plug flow reactors.

Special consideration needs to be given to heterogeneous reactors, in which interaction of the phases is required for the reactions to proceed. In these situations, the rate of reaction may not be the deciding factor in the reactor design. The rate of transport of reactants and/or products from one phase to another can limit the rate at which products are obtained. For example, if reactants cannot get to the surface of a solid catalyst faster than they would react at the surface, then the overall (observed) rate of the process is controlled by this mass transfer step. To improve the rate, the mass transfer must be increased. It would be useless to make changes that would affect only the surface reaction rate. Furthermore, if products do not leave the surface rapidly, they may block reaction sites and thus limit the overall rate. Efficient contacting patterns need to be utilized. Hence, fluidized bed reactors (two-phase backmixed emulator), trickle bed systems (three-phase packed bed emulator), and slurry reactors (three-phase backmixed emulator) have evolved as important bioreactors. They are readily simulated, designed, scaled up, and modified to meet specific contacting demands in cell and tissue culture systems (Shuler and Kargi, 2001).

Flow reactors of all shapes, sizes, and uses are encountered in all walks of life. Examples of interest to bioengineers include the pharmaceutical industry, to produce aspirin, penicillin, and other drugs; the biomass processing industry, to produce alcohol, enzymes and other specialty proteins, and value-added products; and the biotechnologically important tissue and cell culture systems. The type of reactor used depends on the specific application and on the scale desired. The choice is based on a number of factors. The primary ones are the reaction rate (or other rate-limiting process), the product distribution specifications, and the catalyst or other material characteristics, such as chemical and physical stability.

An interesting example of a reactor type useful for a transport rate process is one utilizing impinging jet technology. Here, process intensification concepts are exploited for the general area of microfluidics. The objective is to form nanoparticles via controlled crystallization. Intense mixing is obtained in the unstable supersaturated phase region for a solvent/antisolvent crystallization process, developing a "bottom-up" approach for size control (Panagiotou and Fisher, 2008; Panagiotou et al., 2009). Many pharmaceutical agents can be produced in this manner and/or when coupled with chemical reactions in both miscible and multiphase systems (Baldyga and Bourne, 1999; Johnson and Prud'homme, 2003; Schwarzer and Peukert, 2004; Rabinow, 2004, 2005; Brennen, 2005; Gradl and Peukert, 2009).

Selection of the appropriate reactor type, when living systems are present, requires more thorough discussions since, in these systems, the solid phase may change its dimensions as the reaction proceeds. Particles that increase in size, such as growing cell clusters, can fall out of the reaction zone and alter flow patterns within the vessel. This occurs when gravity overcomes fluid buoyancy and drag forces. In some instances, this biomass growth is the desired product; however, the substances produced by reactions catalyzed by the cellular enzymes are usually the desired products. By reproducing themselves, these cells provide more enzymes and thus productivity increases are possible. Note, however, that there are both advantages and disadvantages of using these whole-cell systems versus the enzymes directly. The cell membrane can be a resistance for transport; the consumption of reactant as a nutrient for cell processes reduces the efficiency of raw material usage; and special precautions must be taken to maintain a healthy environment and thus productivity (Freshney, 2000; Lewis and Colton, 2004; Johnson et al., 2009). The payoff, however, is that the enzymes within their natural environment are generally more active and safer from poisons or other factors that could reduce their effectiveness. Interesting examples are microbial fermentation processes, as discussed earlier. Backmixed flow reactors are used in these applications. They are best suited to maintain the cell line at a particular stage in its life cycle to obtain the desired results. The uniform and constant environment provided for the cells minimizes the adjustments that they need to make concerning nutrient changes, metabolic wastes, and so forth, such that production can proceed at a constant rate. The term "chemostat" is used when referring to backmixed reactors used in biotechnology applications. These are typically tank-type systems with mechanical agitation. All the reactor types discussed earlier, however, are applicable. Recall that mixed flow characteristics can be obtained in tubular reactors if recycle and/or side entry is employed. Thus, an air lift system using a vertical column with recycle and side entry ports is a popular design.

The design, analysis, and simulation of reactors thus becomes an integral part of the bioengineering profession. The study of chemical kinetics, particularly when coupled with complex physical phenomena, such as the transport of heat, mass, and momentum, is required to determine or predict reactor performance. It thus becomes imperative to uncouple and unmask the fundamental phenomenological events in reactors and to subsequently incorporate them in a concerted manner to meet the objectives of specific applications. This need further emphasizes the role played by the physical aspects of reactor behavior in the stability and controllability of the entire process. The following chapters in this book demonstrate the importance of all the concepts presented in this introduction.

References

Baldyga, J. and Bourne, J.R. 1999. *Turbulent Mixing and Chemical Reactions*, John Wiley and Sons, Ltd, Chichester, England.

Bird, R.B., Stewart, W.E., and Lightfoot, E.N. 2002. *Transport Phenomena*, 2nd Edition, John Wiley and Sons, New York.

Brennen, C.E. 2005. *Fundamentals of Multiphase Flow*, Cambridge University Press, New York.

Crank, J. 1956. *The Mathematics of Diffusion*, Oxford University Press, Oxford.

Deen, W.M. 1996. *Analysis of Transport Phenomena*, Oxford Press, New York.

Enderle, J., Blanchard, S., and Bronzino, J.D., Eds. 1999. *Introduction to Biomedical Engineering*, Academic Press, New York.

Fisher, R.J. 1989. Diffusion with immobilization in membranes: Part II, in *Biological and Synthetic Membranes*, Butterfield, A., Ed., Alan R. Liss, Inc., New York, pp. 138–151.

Fisher, R.J. 1999. Compartmental analysis, in *Introduction to Biomedical Engineering*, Academic Press, New York, Chapter 8.

Fournier, R.L. 1999. *Basic Transport Phenomena in Biomedical Engineering*, Taylor & Francis, Philadelphia.

Freshney, R.I. 2000. *Culture of Animal Cells*, 4th Edition, Wiley-Liss, New York.

Gradl, J. and Peukert, W. 2009. Simultaneous 3-D observation of different kinetic sub-processes for precipitation in a T-mixer, *Chem. Eng. Sci.*, 64, 709–720.

Johnson, A.E., Fisher, R.J., Weir, G.C., and Colton, C.K. 2009. Oxygen consumption and diffusion in assemblages of respiring spheres: Performance enhancement of a bioartificial pancreas, *Chem. Eng. Sci.*, 64(22), 4470–4487.

Johnson, B. and Prud'homme, R. 2003. Chemical processing and micro-mixing in confined impinging jets, *AIChE J.*, 49(9), 2264–2282.

Levenspiel, O. 1989. *The Chemical Reactor Omnibook*, Oregon State University Press, Corvallis, OR.

Lewis, A.S. and Colton, C.K. 2004. Tissue engineering for insulin replacement in diabetes, in *Scaffolding in Tissue Engineering*, Ma, P.X. and Elisseeff, J., Eds., Marcel Dekker, New York.

Lightfoot, E.N. 1974. *Transport Phenomena and Living Systems*, John Wiley and Sons, New York.

Panagiotou, T. and Fisher, R.J. 2008. Form Nanoparticles via Controlled Crystallization: A Bottom-up Approach, *Chem. Eng. Prog.*, 10 (Oct.), 33–39.

Panagiotou, T., Mesite, S., and Fisher, R. J. 2009. Production of norfloxacin nano-suspensions using microfluidics reaction technology (MRT) through solvent/anti solvent crystallization, *Ind. Eng. Chem. Res.*, 48(4), 1761–1771.

Rabinow, B. 2004. Nanosuspensions in drug delivery, *Nat. Rev. Drug Discov.*, 3, 785–796.

Rabinow, B. 2005. Pharmacokinetics of drugs administered in nanosuspensions, *Discov. Med.*, 5(25), 74–79.

Rosner, D.E. 1986. *Transport Processes in Chemically Reacting Flow Systems*, Butterworth Publishers, Boston, MA.

Schwarzer, H.C. and Peukert, W. 2004. Tailoring particle size through nanoparticle precipitation, *Chem. Eng. Comm.*, 191, 580–608.

Shuler, M.L. and Kargi F. 2001. *Bioprocess Engineering: Basic Concepts*, 2nd Edition, Prentice-Hall, Englewood Cliffs, NJ.

Editors

Robert A. Peattie earned his BS in biology from Trinity College, Hartford, Connecticut, in 1979 and his PhD in chemical engineering/biomechanics from the Johns Hopkins University in 1988. He has served on the faculties of Tulane University, Trinity College, and Oregon State University, and he is presently an associate research professor in the department of biomedical engineering at Tufts University.

Dr. Peattie is the director of the Tufts Biomechanics-Hemodynamics Laboratory. His research seeks to understand the responses of cells and tissues to applied mechanical challenges in both health and disease, using analytical, computational, and experimental approaches. He has authored over 125 scholarly articles, chapters, and archival contributions, and he has received numerous research and instructional awards, including the international Materialise MIMICS Innovation Award. His teaching interests are centered on physiology and basic medical science, biomechanics, fluid flow and transport, and biofluid mechanics.

Robert J. Fisher is a senior faculty member in the chemical engineering department at the Massachusetts Insitute of Technology and a station director for the David H. Koch School of Chemical Engineering Practice. He is also director of the SABRE Institute for the Engineering Biosciences, a nonprofit research affiliate in collaboration with multiple academic institutions and industrial partners. His current research efforts in nanotechnology include bioprocessing, biomimetics, drug delivery, targeting, imaging, and cell therapy. He has also done extensive consulting for a broad range of industries requiring expertise in stability of transport and reacting systems. As such, he has helped develop innovative process intensification systems and novel applications. One of these collaborations earned a NANO-50 award for advances in impinging jet technology and reaction engineering. Over 200 published documents, including book chapters and patents, have been generated from his efforts covering a broad spectrum of technologies, consistent with a focus on integrated applied mathematics with reaction engineering and transport phenomena. Dr. Fisher's 25-year academic career began at the University of Delaware with the chemical engineering department's Center for Catalytic Science and Technology, after several years with Mobil R&D Corporation. He has received honors and awards for both research and teaching excellence. All his degrees are in chemical engineering: BS/MS from SUNY at Buffalo and PhD from the University of Delaware.

Joseph D. Bronzino earned a BSEE from Worcester Polytechnic Institute, Worcester, Massachusetts, in 1959, a MSEE from the Naval Postgraduate School, Monterey, California, in 1961, and a PhD in electrical engineering from Worcester Polytechnic Institute in 1968. He is presently the Vernon Roosa Professor of Applied Science, an endowed chair at Trinity College, Hartford, Connecticut, and president of the Biomedical Engineering Alliance and Consortium (BEACON), which is a nonprofit organization consisting of academic and medical institutions as well as corporations dedicated to the development of new medical technology. To accomplish this goal, BEACON facilitates collaborative research, industrial partnering, and the development of emerging companies.

Dr. Bronzino is the author of over 200 journal articles and 15 books, including *Technology for Patient Care* (C.V. Mosby, 1977), *Computer Applications for Patient Care* (Addison-Wesley, 1982), *Biomedical Engineering: Basic Concepts and Instrumentation* (PWS Publishing Co., 1986), *Expert Systems: Basic Concepts* (Research Foundation of State University of New York, 1989), *Medical Technology and Society: An Interdisciplinary Perspective* (MIT Press and McGraw-Hill, 1990), *Management of Medical Technology* (Butterworth/Heinemann, 1992), *The Biomedical Engineering Handbook* (CRC Press, 1st edition, 1995; 2nd edition, 2000; 3rd edition, 2006), *Introduction to Biomedical Engineering* (Academic Press, 1st edition, 1999; 2nd edition, 2006), *Biomechanics: Principles and Applications* (CRC Press, 2002), *Biomaterials: Principles and Applications* (CRC Press, 2002), *Tissue Engineering* (CRC Press, 2002), and *Biomedical Imaging* (CRC Press, 2002).

Dr. Bronzino is a fellow of IEEE and the American Institute of Medical and Biological Engineering (AIMBE), an honorary member of the Italian Society of Experimental Biology, past chairman of the Biomedical Engineering Division of the American Society for Engineering Education (ASEE), a charter member of the Connecticut Academy of Science and Engineering (CASE), a charter member of the American College of Clinical Engineering (ACCE), a member of the Association for the Advancement of Medical Instrumentation (AAMI), past president of the IEEE-Engineering in Medicine and Biology Society (EMBS), past chairman of the IEEE Health Care Engineering Policy Committee (HCEPC), and past chairman of the IEEE Technical Policy Council in Washington, DC. He is a member of Eta Kappa Nu, Sigma Xi, and Tau Beta Pi. He is also a recipient of the IEEE Millennium Medal for "his contributions to biomedical engineering research and education" and the Goddard Award from WPI for Outstanding Professional Achievement in 2005. He is presently editor in chief of the Academic Press/Elsevier BME Book Series.

Donald R. Peterson earned his PhD in biomedical engineering and his MS in mechanical engineering at the University of Connecticut. He has been an active member of the Biomedical Engineering (BME) faculty since 1999, offering courses in biomechanics, biodynamics, biofluid mechanics, and ergonomics, and during the past year he has served as BME Graduate Program Committee chair. He is an assistant professor in the School of Medicine, where he is the director of the Biodynamics Laboratory and the Bioengineering Center at the University of Connecticut Health Center. His research involves the modeling of human interactions with existing and developmental devices such as powered and nonpowered tools, dental instruments, computer input devices, musical instruments, sports equipment, and spacesuit and space tool development for NASA. Dr. Peterson has written more than 45 scholarly publications appearing in journals and textbooks.

Contributors

James W. Baish
Department of Biomedical Engineering
Bucknell University
Lewisburg, Pennsylvania

Mandy B. Esch
Department of Biomedical Engineering and
    School of Chemical and Biomolecular
    Engineering
Cornell University
Ithaca, New York

Robert J. Fisher
The SABRE Institute and Department of
    Chemical Engineering
Massachusetts Institute of Technology
Cambridge, Massachusetts

Bingmei M. Fu
Department of Biomedical Engineering
The City College of the City University of New York
New York, New York

Donald P. Gaver III
Department of Biomedical Engineering
Tulane University
New Orleans, Louisiana

Sarina G. Harris
Department of Biomedical Engineering and
    School of Chemical and Biomolecular
    Engineering
Cornell University
Ithaca, New York

Anne-Marie Jacob
Department of Biomedical Engineering
Tulane University
New Orleans, Louisiana

Xinran Li
Department of Biomedical Engineering and School
    of Chemical and Biomolecular Engineering
Cornell University
Ithaca, New York

E. N. Lightfoot
Department of Chemical and Biological Engineering
University of Wisconsin
Madison, Wisconsin

David Martin
Department of Biomedical Engineering
Tulane University
New Orleans, Louisiana

Robert A. Peattie
Department of Biomedical Engineering
Tufts University
Medford, Massachusetts

Yuchen Qiu
Cordis Corporation
Bridgewater, New Jersey

W. Mark Saltzman
Department of Biomedical Engineering
Yale University
New Haven, Connecticut

Michael L. Shuler
Department of Biomedical Engineering and School
    of Chemical and Biomolecular Engineering
Cornell University
Ithaca, New York

John M. Tarbell
Department of Biomedical Engineering
The City College of the City University of New York
New York, New York

1

Biomimetic Systems: Concepts, Design, and Emulation

1.1 Concepts of Biomimicry... 1-2
Morphology and Properties Development • Molecular Engineering of Thin Films and Nanocapsules • Biotechnology, Bioreaction Engineering, and Systems Development

1.2 Biomimicry and Tissue Engineering... 1-4
Integrated Systems • Blood–Brain Barrier • Vascular System • Implants

1.3 Biomimetic Membranes for Ion Transport 1-8
Active Transport Biomimetics • Mechanism for Facilitated Diffusion in Fixed Carrier Membranes • Jumping Mechanism in Immobilized Liquid Membranes

1.4 Assessing Mass Transfer Resistances in Biomimetic Reactors 1-11
Uncoupling Resistances • Use in Physiologically Based Pharmacokinetics Models and Cell Culture Analog Systems

1.5 Electro-Enzymatic Membrane Reactors as Electron Transfer Chain Biomimetics.. 1-13
Mimicry of *In Vivo* Coenzyme Regeneration Processes • Electro-Enzymatic Production of Lactate from Pyruvate

References... 1-14

Robert J. Fisher
The SABRE Institute and Massachusetts Institute of Technology

Humans have always been fascinated by the phenomenological events, both biological and physical in nature, that are revealed to us by our environment. Our innate curiosity drives us to study these observations and understand the fundamental basis of the mechanisms involved. Practical outcomes are the development of predictive capabilities of occurrence and the control of these events and their subsequent consequences; our safety and comfort being major incentives. Furthermore, we wish to design processes that mimic the beneficial aspects associated with their natural counterparts. Experience has taught us that these natural processes are complex and durable and that adaptability with multifunctionality is a must for biological systems to survive. Evolution, aiding these living systems to adapt to new environmental challenges, occurs at the molecular scale. Our need to be molecular scientists and engineers is thus apparent. Knowledge of the molecular building blocks used in the architectural configurations of both living and nonliving systems, along with an understanding of their design and the processes used for implementation, is essential for control and utilization. The ability to mimic demonstrates a sufficient knowledge base to design systems requiring controlled functionality. To perfect this approach, a series of sensor/reporter systems must be available. A particularly attractive feature of living systems is their unique ability to diagnose and repair localized damages through a continuously distributed sensor network with inter- and intracellular communication capabilities. Mimicry of these networks is an integral component of many emerging research thrust areas, in particular, tissue engineering. Significant emphasis has been toward the development of intelligent membranes, specifically using sensor/reporter technology. A successful approach has been to couple transformation and separation technologies with detection and control systems.

The concept of intelligent barriers and substrates, such as membranes, arises from the coupling of this sensor/reporter technology with controlled chemistry and reaction engineering, selective transport phenomena, and innovative systems design. Engineered membrane mimetics are required to respond and adapt to environmental stresses, whether intentionally imposed or stochastic in nature. These intelligent membranes may take the form of polymeric films, composite materials, ceramics, supported liquid membranes, or as laminates. Their important feature is specific chemical functionality, engineered to provide selective transport, structural integrity, controlled stability and release, and sensor/reporter capabilities. Applications as active transport and electron transfer chain mimics, and their use in studying the consequences of environmental stresses on enzymatic functions, create valuable insights into cellular mechanisms.

Advanced materials, designed through knowledge gained from analysis of biological systems, have been instrumental in the progression and success of many tissue engineering applications. Their biocompatibility, multifunctionality, and physiochemical properties are essential attributes. When incorporated with living cells, for example, in organ/tissue constructs, integrated systems biology behavioral mimicry can be accomplished. This is particularly useful for drug efficacy and toxicity screening tests and thus minimizing the use of animals for these studies. For example, having an effective blood–brain barrier (BBB) mimetic, a realistic blood substitute, and cell culture analogs (CCAs) for the various organs needed for a useful animal surrogate system promotes more rapid development of therapeutic drugs. These biomimetic studies influence all phases, that is, the design, development, and delivery characteristics, of this effort. The design, applicability, and performance of these systems are briefly discussed in subsequent sections in this chapter and in greater depth in other chapters throughout this handbook.

1.1 Concepts of Biomimicry

Discoveries that have emerged from a wide spectrum of disciplines, ranging from biotechnology and genetics to polymer and molecular engineering, are extending the design and manufacturing possibilities for mimetic systems that were once incomprehensible. Understanding of the fundamental concepts inherent in natural processes has led to a broad spectrum of new processes and materials modeled on these systems (Srinivasan et al., 1991). Natural processes, such as active transport systems functioning in living systems, have been successfully mimicked and useful applications in other fields, such as pollution prevention, have been demonstrated (Thoresen and Fisher, 1995). Understanding the mechanisms at the molecular level, which living systems utilize, is needed before success at the macroscale can be assured. Multifunctionality, hierarchical organization, adaptability, reliability, self-regulation, and reparability are the key elements that living systems rely upon and that we must mimic to develop "intelligent systems." Successes to date have been based on the use of techniques associated with research advances made in areas such as molecular engineering of thin films and nanocapsules, neural networks, reporter/sensor technology, morphology and properties developments in polymer blends, transport phenomena in stationary and reacting flow systems, cell culture and immobilization technologies, controlled release and stability mechanisms for therapeutic agents, and environmental stress analyses. A few examples in the bioengineering field are molecular design of supported liquid membranes to mimic active transport of ions; noninvasive sensors to monitor *in vivo* glucose concentrations; detection of microbial contamination by bioluminescence; immunomagnetic capture of pathogens; improved encapsulation systems; carrier molecules for targeting, imaging, and controlled release of pharmaceutics; *in situ* regeneration of coenzymes electrochemically; and measurement of transport and failure mechanisms in "smart" composites. All these successes were accomplished through interdisciplinary approaches, essentially using three major impact themes: morphology and properties development; molecular engineering of thin films and nanocapsules; and biotechnology, bioreaction engineering, and systems development.

1.1.1 Morphology and Properties Development

Polymer blends are a major focus of this research area (Weiss et al., 1995). Technology, however, has outpaced a detailed understanding of many facets of this science, which impedes the development and application of new materials. The purpose of this research is to develop the fundamental science that influences the phase behavior, phase architecture, and morphology and interfacial properties of polymer blends. The main medical areas where polymers and composites (polymeric) have found wide use are artificial organs, the cardiovascular system, orthopedics, dental sciences, ophthalmology, and drug delivery systems. Success has been related to the wide range of mechanical properties, transformation processes (shape possibilities), and low production costs. The limitation has been their interaction with living tissue. To overcome the biological deficiencies of synthetic polymers and to enhance the mechanical characteristics, a class of bioartificial polymeric materials has been introduced based on blends, composites, and interpenetrating polymer networks of both synthetic and biological polymers. Preparations from biopolymers such as fibrin, collagen, and hyaluronic acid with synthetic polymers such as polyurethane, poly(acrylic acid), and poly(vinyl alcohol) are available (Giusti et al., 1993; Luo and Prestwich, 2001).

1.1.2 Molecular Engineering of Thin Films and Nanocapsules

The focus in the thin-film research impact area is to develop a fundamental understanding of how morphology can be controlled in (1) organic thin-film composites prepared by Langmuir–Blodgett (LB) monolayer and multilayer techniques and (2) the molecular design of membrane systems using ionomers and selected supported liquids. Controlled structures of this nature will find immediate application in several aspects of smart materials development, particularly in microsensors.

The ability to form nanosized particles and/or emulsions that encapsulate active ingredients is an essential skill applicable to many facets of the engineering biosciences. Nanotechnologies have a major impact on drug delivery, molecular targeting, medical imaging, biosensor development, and in the cosmetic, personal care products, and nutraceutics industries, to mention only a few. New techniques utilize high shear fields to obtain particle sizes in the range of 50–100 nm; about the size of the turbulent eddies developed. Stable emulsions can be formed with conventional mixing equipment where high shear elongation flow fields are generated near the tip of high-speed blades, but only in the range 500 nm and larger. High shear stresses can also be generated by forcing the components of the microemulsion to flow through a microporous material. The resultant solution contains average particle sizes as small as 50 nm. Units that incorporate jet impingement on a solid surface or with another jet also perform in this size range. Molecular self-assembly systems using novel biocompatible synthetic polymer compounds are also under development. These have been used successfully to encapsulate chemotherapeutic drugs and perflourocarbons (PFC) (Kumar et al., 2005). The high oxygen solubility of PFCs makes them attractive as blood surrogates and also as additives in immuno-isolation tissue encapsulation systems to enhance gas transport (Johnson et al., 2009). Using nanoencapsulation techniques, these compounds can be dispersed throughout microencapsulating matrices and/or in tissue extracellular matrix (ECM) scaffold systems.

Surfaces, interfaces, and microstructures play an important role in many research frontiers. Exploration of structural property relationships at the atomic and molecular level, investigating elementary chemical and physical transformations occurring at phase boundaries, applying modern theoretical methods for predicting chemical dynamics at surfaces, and integration of this knowledge into models that can be used in process design and evaluation are within the realm of surface and interfacial engineering. These concepts are also important in drug design and delivery. Both crystal size and its morphology impact type and rate of uptake, its partitioning characteristics, and efficacy. Through proper selection of processing conditions, the method used to obtain the supersaturation state and its magnitude, and use of surface active agents, one can control product properties. A particular polymorph can be formed by controlling/directing the self-assembly mechanisms. Furthermore, stability with respect to both crystal size and morphology is possible (Rabinow, 2004, 2005; Schwarzer and Peukert, 2004; Panagiotou and Fisher, 2008; Panagiotou et al., 2009).

The control of surface functionality by proper selection of the composition of the LB films and/or the self-assembling (amphiphatic) molecular systems can mimic many functions of a biologically active membrane. An informative comparison is that between inverted erythrocyte ghosts (Dinno et al., 1991; Matthews et al., 1993) and their synthetic mimics when environmental stresses are imposed on both systems. These model systems can assist in mechanistic studies to understand the functional alterations that result from ultrasound, EM fields, and UV radiation. The behavior of carrier molecules and receptor site functionality must be mimicked properly along with simulating disturbances in the proton motive force (PMF) of viable cells. Use of ion/electron transport ionomers in membrane–catalyst preparations is beneficial for programs such as electro-enzymatic synthesis and metabolic pathway emulation (Fisher et al., 2000; Chen et al., 2004). Development of new membranes used in artificial organs and advances in micelle reaction systems have resulted from these efforts.

1.1.3 Biotechnology, Bioreaction Engineering, and Systems Development

Focus for this research area is on: (1) sensor/receptor reporter systems and detection methods; (2) transport processes in biological and synthetic membranes; (3) biomedical and bioconversion process development; and (4) smart film/intelligent barrier systems. These topics require coupling with the previously discussed areas and the use of biochemical reaction engineering techniques. Included in all of these areas is the concept of metabolic engineering; the modification of the metabolism of organisms to produce useful products. Extensive research in bioconversion processes is currently being directed to producing important pharmaceutics. Expanded efforts are also needed in the field of cell and tissue engineering, that is, the manipulation or reconstruction of cell and tissue function using molecular approaches (Johnson et al., 2009).

1.2 Biomimicry and Tissue Engineering

Before we can develop useful *ex vivo* and *in vitro* systems for the numerous applications in tissue engineering, we must have an appreciation of cellular function *in vivo*. Knowledge of the tissue microenvironment and communication with other organs is essential. The key questions that must therefore be addressed in the realm of tissue engineering are thus, how can tissue function be built, reconstructed, and/or modified? To answer these, we develop a standard approach based on the following axioms (Palsson, 2000): (1) in organogenesis and wound healing, proper cellular communications, with respect to each other's activities, are of paramount concern since a systematic and regulated response is required from all participating cells; (2) the function of fully formed organs is strongly dependent on the coordinated function of multiple cell types with tissue function based on multicellular aggregates; (3) the functionality of an individual cell is strongly affected by its microenvironment (within 100 μm of the cell, that is, the characteristic length scale); (4) this microenvironment is further characterized by (i) neighboring cells, that is, cell–cell contact and presence of molecular signals (soluble growth factors, signal transduction, trafficking, etc.), (ii) transport processes and physical interactions with the ECM, and (iii) the local geometry, in particular its effects on microcirculation. The importance of the microcirculation is that it connects all microenvironments to the whole-body environment. Most metabolically active cells in the body are located within a few hundred micrometers from a capillary.

This high degree of vascularization is necessary to provide the perfusion environment that connects every cell to a source and sink for respiratory gases, a source of nutrients from the small intestine, the hormones from the pancreas, liver, and glandular system, clearance of waste products via the kidneys and liver, delivery of immune system respondents, and so forth. The engineering of these functions *ex vivo* is the domain of bioreactor design, a topic discussed briefly in the introductory comments to this book and also in Chapters 4 and 7. These cell culture devices must appropriately simulate and provide these macroenvironmental functions while respecting the need for the formation of microenvironments. Consequently, they must possess perfusion characteristics that allow for uniformity down to the 100 μm length scale. These are stringent design requirements that must be addressed with a high priority to properly account for the role of neighboring cells, the ECM, cyto-/chemokine and hormone trafficking, geometry, the dynamics of respiration, and the transport of nutrients and metabolic by-products for each tissue system considered. These dynamic, chemical, and geometric variables must be duplicated as accurately as possible to achieve proper mimicry. Since this is a difficult task, a significant portion of Chapter 7 in this book is devoted to developing methods to describe the microenvironment. Using the tools discussed there we can develop systems to control microenvironments for *in vivo*, *ex vivo*, or *in vitro* applications.

The approach taken here to achieve the desired microenvironments is through use of novel membrane systems. They are designed to possess unique features for the specific application of interest and in many cases to exhibit stimulant/response characteristics. These so-called "intelligent" or "smart" membranes are the result of biomimicry, that is, having biomimetic features. Through functionalized membranes, typically in concerted assemblies, these systems respond to external stresses (chemical and/or physical in nature) to eliminate the threat either by altering stress characteristics or by modifying and/or protecting the cell/tissue microenvironment. An example is a microencapsulation motif for beta cell islet clusters to perform as a pancreas. This system uses multiple membrane materials, each with its unique characteristics and performance requirements, coupled with nanospheres dispersed throughout the matrix, which contain additional materials for enhanced transport and/or barrier properties and respond to specific stimuli (Galletti et al., 2000; Lewis and Colton, 2004; Peattie et al., 2004; Johnson et al., 2009).

The communication of every cell with its immediate environment and other tissues is a key requisite for successful tissue function. This need establishes important spatial-temporal characteristics and a significant signaling/information processing network (Lauffenburger and Linderman, 1993). Understanding this network and the information it contains is what the tissue engineer wishes to express and manage. For example, to stimulate the beginning of a specific cellular process appropriate signals to the nucleus are delivered at the cell membrane and transmitted through the cytoplasm by a variety of signal transduction mechanisms. Some signals are delivered by soluble growth factors that may originate from the circulating blood or from neighboring cells. The signal networking process is initiated after these molecules bind to selective receptors. The microenvironment is also characterized by cellular composition, the ECM, molecular dynamics (nutrients, metabolic waste products, and respiratory gases traffic in and out of the microenvironment in a highly dynamic manner), and local geometry (size scale of approximately 100 μm). Each of these can also provide the cell with important signals (dependent upon a characteristic time and length scale) to initiate specific cell functions for the tissue system to perform in a coordinated manner. If this arrangement is disrupted, cells that are unable to provide tissue function are obtained. Further discussions on this topic are presented in other chapters in this handbook devoted to cellular communications.

1.2.1 Integrated Systems

The interactions brought about by communications between tissue microenvironments and the whole-body system, via the vascular network, provide a basis for the systems biology approach taken to understand the performance differences observed *in vivo* versus *in vitro*. The response of one tissue system to changes in another (due to signals generated, such as metabolic products or hormones) must be properly mimicked by coupling individual CCA systems through a series of microbioreactors if whole-body *in vivo* responses are to be meaningfully predicted. The need for microscale reactors is obvious when we consider the limited amount of tissue/cells available for these *in vitro* studies. This is particularly true when dealing with the pancreatic system where intact islets must be used (vs. individual beta cells) for induced insulin production by glucose stimulation (however, see Johnson et al. (2009) for a promising new approach). Their supply is extremely limited and maintaining viability and functionality is quite complex since the islet clusters *in vivo* are highly vascularized and this feature is difficult to maintain in preservation protocols or reproduce in mimetic systems. Therefore, the time scale for their usefulness is limited. Furthermore, one needs to minimize the amount of serum used (the communication fluid flowing through and between these biomimetic reactors), since in many cases, the serum obtained from actual patients must be used for proper mimicry. An animal surrogate system, primarily for drug toxicity studies, is currently being developed using this CCA concept (Shuler et al., 1996). A general CCA system is one of three topics selected to illustrate these system interaction concepts in subsequent subsections in this chapter and others in this handbook. Another is associated with the use of compartmental analysis in understanding the distribution and fate of molecular species, particularly pharmaceutics, and the third is the need for facilitated transport across the BBB, due to its complexities, when these species are introduced into the whole body by systemic administration.

Compartmental analysis and modeling was first formalized in the context of isotropic tracer kinetics to determine distribution parameters for fluid-borne species in both living and inert systems, particularly useful for determining flow patterns in reactors and tissue uptake parameters (Fisher, 2000). Over time, it has evolved and grown as a formal body of theory (Michaels, 1988; Fisher, 2000). Models developed using compartmental analysis techniques are a class of dynamic, that is, differential equation, models derived from mass balance considerations. These compartmental models are widely used for quantitative analysis of the kinetics of "materials" in physiologic systems. These materials can be either exogenous, such as a drug or a tracer, or endogenous, such as a reactant (substrate) or a hormone. Kinetics include processes such as production, distribution, transport, utilization, and substrate–hormone control interactions.

1.2.2 Blood–Brain Barrier

Many drugs, particularly water-soluble or high-molecular-weight compounds, do not enter the brain following traditional systemic administration methods because their permeation rate through blood capillaries is very slow. This BBB severely limits the number of drugs that are candidates for treating brain disease. New strategies for increasing the permeability of brain capillaries to drugs are constantly being tested and are discussed elsewhere in other chapters of this handbook. A seemingly effective technique, of particular interest for this section is to utilize specific nutrient transport systems in brain capillaries to facilitate drug transport. For example, certain metabolic precursors are transported across endothelial cells by the neutral amino acid transport system and therefore, analog compounds could be used as both chaperones and targeting species. Also, direct delivery into the brain tissue by infusion, implantation of a drug-releasing matrix, and transplantation of drug-secreting cells are being considered. These approaches provide sustained drug delivery that can be confined to specific sites, localizing therapy to a given region. As they provide a localized and continuous source of active drug molecules, the total drug dosage can be less than with systemic administration. Polymeric implants, for controlled drug release, can also be designed to protect unreleased drug from degradation in the body and to permit localization of extremely high doses at precisely defined locations in the brain. Infusion systems require periodic refilling. This usually requires the drug to be stored in a liquid reservoir at body temperature and therefore many drugs are not suitable for this application since they are not stable under these conditions.

Coupling of these approaches, using nanosphere technologies to entrap the drug (sometimes modified for enhanced encapsulation stability) along with surface modifications of the spheres for specific targeting, has a higher probability of success to enhance transport into the brain. Proof-of-concept experiments can be conducted using a valid BBB model as an effective biomimetic. An *in vitro* coculture system comprised of porcine brain capillary endothelial cells (BCEC) with porcine astrocytes is widely accepted as a valid BBB model. Using standard cell culturing techniques, the astrocytes are seeded on the bottom of permeable membrane filters with BCEC seated on the top. This configuration permits communication between the two cell lines without disruption of the endothelial cell monolayer. The filters are suspended in a chamber of fluid such that an upper chamber is formed analogous to the lumen of a brain capillary blood vessel. BBB permeability is determined from the measurement of transendothelial electrical resistance (TEER), a standard technique that measures the cell layer's ability to resist the passage of a low electrical current. It essentially represents the passages of small ions and is the most sensitive measure of BBB integrity. The nanospheres can be used to encapsulate the radioactive-labeled drug and tested for their toxicity to the BBB using TEER and noting if loss of barrier properties are observed. Inulin (5200 Da) is used as a marker species to represent potential pharmaceutical drug candidates and when used without nanosphere encapsulation, provides a reasonable control. For example, its transport across the BBB is quite slow; less than 2% after 4 h of exposure. In proof-of-concept experiments, for the same time period and drug concentrations, more than 16% of the inulin within the nanospheres crossed the BBB (Kumar et al., 2005). Although mechanistic details are lacking, this greater than eightfold increase in rate represents a dramatic increase and supports the premise that nanosphere encapsulation can facilitate drug delivery across the BBB and further illustrates the usefulness of this BBB biomimetic system.

1.2.3 Vascular System

Nutrient supply and gas exchange can become limiting in high cell density situation, as in tissue emulation *in vitro*, due to lack of an effective vasculature mimetic system to provide *in vivo* perfusion conditions. Many different system configurations/designs have been considered to overcome these deficiencies, such as cellulose and gel-foam sponge matrix materials, filter-well inserts, and mimetic membranes in novel bioreactor systems (Freshney, 2000). Of particular interest here is the use of synthetic polymer capillary fibers (hollow fibers) in perfusion chambers where they can support cell growth on their outer surfaces and are gas and nutrient permeable. Medium, saturated with 5% CO_2 in air, is pumped through the lumen of the capillary fibers (in a bundle configuration) and cells attached and growing on the outer surface of the fibers, fed by diffusion from the perfusate, can reach tissue-like cell densities. Different polymers and ultrafiltration properties provide molecular cut-offs at 10–100 kDa, regulating macromolecule diffusion. It is now possible for the cells to behave as they would *in vivo*. For example, in such cultures, choriocarcinoma cells release more human chorionic gonadotrophin than they would in conventional monolayer culture and colonic carcinoma cells produce elevated levels of carcinoembryonic antigen (Freshney, 2000). Unfortunately, sampling cells and determining cell density are difficult from these commercially available chambers. New configurations are presently being designed and tested to overcome these limitations and are discussed in Chapters 4 and 7 in this book.

1.2.4 Implants

The transport of mass to and within a tissue is determined primarily by convection and diffusion processes that occur throughout the whole-body system. The design of systems, for example, in cellular therapy, must consider methods to promote this integrated process and not just deal with the transport issues of the device itself. An encapsulated tissue system implant must develop an enhanced localized vasculature. This may be accomplished by (1) recruiting vessels from the preexisting network of the host vasculature and/or (2) stimulating new vessel growth resulting from an angiogenic response of host vessels to the implant (Jain, 1994; Peattie et al., 2004). Therefore, when considering implantation of encapsulated tissue/cells, it would be prudent to design the implant to have this biomimetic characteristic; to elicit an angiogenic response from a component of and/or in the matrix itself. For example, it is known that hyaluronic hydrogels can be synthesized to be biodegradable and that these degradation products stimulate microvessel growth. Also, any biocompatible matrix could be loaded with cytokines that would diffuse out on their own and/or be released via degradation mechanisms from microcapsules dispersed throughout the ECM. A recent study (Peattie et al., 2004) demonstrated these facts and identified synergistic behaviors. In summary, cross-linked hyaluronic acid (HA) hydrogels were evaluated for their ability to elicit new microvessel growth *in vivo* when loaded with one of two cytokines, vascular endothelial growth factor (VEGF) or basic fibroblast growth factor (bFGF). HA film samples were surgically implanted in the ear pinnas of mice, and the samples retrieved 7 or 14 days postimplantation. Histologic analysis showed that all groups receiving an implant demonstrated significantly more microvessel density than control groups undergoing surgery but receiving no implant. Moreover, aqueous administration of either growth factor produced substantially more vessel growth than an HA implant with no cytokine. However, the most striking result obtained was a dramatic synergistic interaction between HA and VEGF. New vessel growth was quantified by a metric developed during that study; that is, a dimensionless neovascularization index (NI). This index is defined to represent the number of additional vessels present postimplant in a treatment group, minus the additional number due to the surgical procedure alone, normalized by the contralateral count. Presentation of VEGF in cross-linked HA generated vessel density of NI = 6.7 at day 14. This was more than twice the effect of the sum of HA alone (NI = 1.8) plus VEGF alone (NI = 1.3). This was twice the vessel density generated by coaddition of HA and bFGF (NI = 3.4). New therapeutic approaches for numerous pathologies could be notably enhanced by this localized, synergistic angiogenic response produced by the release of VEGF from cross-linked HA films.

1.3 Biomimetic Membranes for Ion Transport

Cells must take nutrients from their extracellular environment to grow and/or maintain metabolic activity. The selectivity and rate that these molecular species enter can be important in regulatory processes. The mechanisms involved depend upon the size of the molecules to be transported across the cell membrane. These biological membranes consist of a continuous double layer of molecules creating a hydrophobic interspace in which various membrane proteins are imbedded. Individual lipid molecules are able to diffuse rapidly within their own monolayer; however, they rarely "flip-flop" spontaneously between these two monolayers. These molecules are amphoteric and assemble spontaneously into bilayers when placed in water. Sealed compartments are thus formed, which reseal if torn.

The topic of membrane transport is discussed in detail in many texts (Alberts et al., 1989; Freshney, 2000). The following discussion is limited to membrane transport of small molecules, hence excluding macromolecules such as polypeptides, polysaccharides, and polynucleotides. The lipid bilayer is a highly impermeable barrier to most polar molecules and thus prevents the loss of the water-soluble contents of the cell interior. Consequently, cells have developed special means to transport these species across their membranes. Specialized transmembrane proteins accomplish this, each responsible for the transfer of a specific molecule or group of closely related molecules. The mechanism can be either energy independent, as in passive and facilitated diffusion, or energy dependent, as in active transport and group translocation.

In passive diffusion, molecules are transported with (or "down") a concentration gradient that is thermodynamically favorable and can occur spontaneously. Facilitated diffusion utilizes a carrier molecule, imbedded in the membrane, that can combine specifically and reversibly with the molecule to be transported. This carrier protein undergoes conformational changes when the target molecule binds and again when it releases that molecule on the transverse side of the membrane. This binding is dependent on favorable thermodynamics related to the concentration of free versus bound species. An equilibrium is established, as in a Langmuir isotherm, on both sides of the membrane. Thus, the rate of transport is proportional to concentration differences maintained on each side of the membrane and the direction of flow is down this gradient. Active transport is similar to facilitated transport in that a carrier protein is necessary; however, it occurs against (up) a concentration gradient, which is thermodynamically unfavorable and thus requires energy. Group translocation requires chemical modification of the substance during the process of transport. This conversion process traps the molecule on a specific side of the membrane due to its asymmetric nature and the essential irreversibility of the transformation. These complexities lead to difficulties in mimicry; thus, research in this area is slow in developing.

Several energy sources are possible for active transport, including electrostatic or pH gradients of the PMF, secondary gradients derived from the PMF by other active transport systems, and by the hydrolysis of ATP. The development of these ion gradients enables the cell to store potential energy in the form of these gradients.

It is essential to realize that simple synthetic lipid bilayers, that is, protein-free, can mimic only passive diffusion processes since they are impermeable to ions but freely permeable to water. Thermodynamically, virtually any molecule should diffuse across a protein-free, synthetic lipid bilayer down its concentration gradient. However, it is the rate of diffusion that is of concern, which is highly dependent upon the size of the molecule and its relative solubility in oil (i.e., the hydrophobic interior of the bilayer). Consequently, small nonpolar molecules such as O_2 readily diffuse. If small enough, uncharged polar molecules such as CO_2, ethanol, and urea can diffuse rapidly, whereas glycerol is more difficult and glucose is essentially excluded. Water, because it has such a small volume and is uncharged, diffuses rapidly even though it is polar and relatively insoluble in the hydrophobic phase of the bylayer. Charged particles, on the other hand, no matter how small, such as Na^+ and K^+, are essentially excluded. This is due to the charge and the high degree of hydration preventing them from entering the hydrocarbon phase. Quantitatively, water permeates the bilayer at a rate 10^3 faster than urea, 10^6 faster than glucose, and 10^9 faster than small ions such as K^+. Thus, only nonpolar molecules and small uncharged polar molecules can cross the cellular lipid membrane directly by simple (passive) diffusion; others require specific membrane transport proteins, as either carriers or channels. Synthetic membranes can be designed for specific biomedical applications that can mimic the transport processes discussed earlier. Membrane selectivity and transport are enhanced with the aid of highly selective complexing agents, impregnated as either fixed site or mobile carriers. To use these membranes to their full potential, the mechanism of this diffusion needs to be thoroughly understood. The following subsections provide some insight for this behavior.

1.3.1 Active Transport Biomimetics

Extensive theoretical and experimental work has previously been reported for supported liquid membrane systems (SLMS) as effective mimics of active transport of ions (Cussler et al., 1989; Kalachev et al., 1992; Thoresen and Fisher, 1995; Stockton and Fisher, 1998). This was successfully demonstrated using di-(2-ethyl hexyl)-phosphoric acid as the mobile carrier dissolved in *n*-dodecane, supported in various inert hydrophobic microporous matrices (e.g., polypropylene), with copper and nickel ions as the transported species. The results showed that a pH differential between the aqueous feed and strip streams, separated by the SLMS, mimics the PMF required for the emulated active transport process that occurred. The model for transport in an SLMS is represented by a five-step resistance-in-series approach, as follows: (1) diffusion of the ion through a hydrodynamic boundary layer; (2) desolvation of the ion, where it expels the water molecules in its coordination sphere and enters the organic phase via ion exchange with the mobile carrier at the feed/membrane interface; (3) diffusion of the ion–carrier complex across the SLMS to the strip/membrane interface; (4) solvation of the ion as it enters the aqueous strip solution via ion exchange; and (5) transport of the ion through the hydrodynamic boundary layer to the bulk stripping solution. A local Peclet number is used to characterize the hydrodynamics and the mass transfer occurring at each fluid/SLMS interface. The SLMS itself is modeled as a heterogeneous surface with mass transfer and reaction occurring only at active sites; in this case, the transverse pores. Long-term stability and toxicity problems limit their application (as configured above) in the biomedical arena. Use in combination with fixed-site carrier membranes as entrapping barriers has great potential and is an active research area. Some success has been obtained using (1) reticulated vitreous carbon as the support matrix and nafion, for the thin-film "active barrier" and (2) an ethylene–acrylic acid ionomer, utilizing the carboxylic acid groups as the fixed-site carriers. The most probable design for biomedical applications appears to be a laminate composite system that incorporates less toxic SLMSs and highly selective molecularly engineered thin-film entrapping membranes. Use of fixed-site carrier membranes in these innovative designs requires knowledge of transport characteristics. Cussler et al. (1989) have theoretically predicted a jumping mechanism for these systems. Kalachev et al. (1992) have shown that this mechanism can also occur in an SLMS at certain carrier concentrations. This mechanism allows for more efficient transport than common facilitated diffusion. Stability over time and a larger range of carrier concentrations where jumping occurs make fixed carrier membranes attractive for biomedical applications. A brief discussion of these jumping mechanisms follows.

1.3.2 Mechanism for Facilitated Diffusion in Fixed Carrier Membranes

A theory for the mechanism of diffusion through a membrane using a fixed carrier covalently bound to the solid matrix was developed previously (Cussler et al., 1989). The concept is that the solute molecule jumps from one carrier to the next in sequence. Facilitated diffusion can occur only if these "chained" carriers are reasonably close to each other and have some limited mobility. The advantages of using a chained carrier in a solid matrix versus a mobile carrier in a liquid membrane are that the stability is improved; there is no potential for solvent loss from the system, and the transport may actually be enhanced. Their theory is compared to that for the mobile carriers in the SLMS. For the fixed carrier (chained) system, the assumptions of fast reactions and that they take place only at the interface, are also used. The major difference is that the complex formed cannot diffuse across the membrane since the carrier is covalently bound to the polymer chain in the membrane. Although the complex does not diffuse in the classical random walk concept, it can "jiggle" around its equilibrium position. This movement can bring it into contact range with an uncomplexed carrier also "jiggling," and result in a reversible interaction typical to normal receptor/ligand surface motion. It is assumed that no uncomplexed solute can pass through the membrane; it would be immobilized and taken from the diffusion process. The transport process that is operable is best explained by viewing the chained carrier membrane as a lamella structure where each layer is of thickness L. Every carrier can move a distance X around its neutral position and is a length L away from its neighbors. Diffusion can occur only over the distance X. Therefore, there is a specific concentration where a solute flux is first detected, termed percolation threshold, occurring when $L = X$. This threshold concentration is estimated as $C = 1/(L^3 N_a)$, where C is the average concentration, L is the distance between carrier molecules, and N_a is Avogadro's number. In summary, the mechanism is that of intramolecular diffusion; each chained carrier having limited mobility within the membrane. A carrier at the fluid membrane interface reacts with the species to be transported and subsequently comes in contact with an uncomplexed carrier and reacts with it, repeating this transfer process across the entire width of the membrane.

1.3.3 Jumping Mechanism in Immobilized Liquid Membranes

Facilitated diffusion was studied in immobilized liquid membranes using a system composed of a microporous nitrocellulose film impregnated with tri-*n*-octylamine (TOA) in *n*-decane (Kalachev et al., 1992).

Experiments were monitored by measuring the conductivity of the feed and strip streams. The transport of ions (cobalt and iron) from an acidic feed (HCl) to a basic strip solution (NH_4OH) was accomplished. Their results suggest that there are three distinct transport regimes operable in the membrane. The first occurs at short times and exhibits very little ion transport. This initial time is termed the ion penetration time and is simply the transport time across the membrane. At long times, a rapid increase in indiscriminate transport is observed. At this critical time and beyond, there are stability problems; that is, loss of solvent from the pores leading to the degradation of the membrane and the formation of channels that compromise the ion selective nature of the system and its barrier properties.

Recall that their experiments were for a selective transport with (not against) the ion gradient. It is only in the intermediate time regime that actual facilitated transport occurs. In this second region, experiments were conducted using a cobalt feed solution for various times and carrier concentrations; all experiments showed a peak in flux. The velocity of the transported species can be obtained from these results and the penetration time versus carrier concentration is available. At the threshold carrier concentration, these researchers claim that the mechanism of transport is by jumping as proposed earlier for fixed-site carriers. The carrier molecules are now close enough to participate in a "bucket brigade" transport mechanism. The carrier molecules use local mobility, made possible by a low viscous solution of *n*-decane, to oscillate, passing the transported species from one to another. This motion results in faster transport than common facilitated transport, which relies on the random walk concept and occurs at lower TOA concentrations. It is in this low concentration region that the carrier molecules are too far apart to participate in the jumping scheme. At higher concentrations, well above the threshold value, the increased viscosity interferes with carrier mobility; the jumping is less direct or does not occur because of the increased bonding sites and hence removal of the species from the transport process.

1.4 Assessing Mass Transfer Resistances in Biomimetic Reactors

1.4.1 Uncoupling Resistances

Characterization of mass transfer limitations in biomimetic reactors is essential when designing and evaluating their performance. When used in CCA systems, the proper mimicry of the role of intrinsic kinetics and transport phenomena cannot be overemphasized. Lack of the desired similitude will negate the credibility of the phenomenological observations as pertaining to toxicity and/or pharmaceutical efficacy. The systems must be designed to allow manipulation, and thus control, of all interfacial events. The majority of material transfer studies for gaseous substrates are based on the assumption that the primary resistance is at the gas/liquid interface. Studies examining the use of hollow fiber membranes to enhance gas/liquid transport have been successfully conducted (Grasso et al., 1995). The liquid/cell interfacial resistance is thus uncoupled from that of the gas/liquid interface and they can now be examined separately to evaluate their potential impacts. A reduction in the mean velocity gradient, while maintaining a constant substrate flux into the liquid, resulted in a shift in the limiting resistance from the gas/liquid to the liquid/cell interface. This shift manifested itself as an increase in the Monad apparent half-saturation constant for the chemo-autotrophic methanogenic microbial system selected as a convenient analog. The result of these studies significantly influences the design and/or evaluation of reactors used in the biomedical engineering (BME) research area, especially for the animal surrogate or CCA systems. Although a reactor can be considered as well mixed based on spatial invariance in cell density, it was demonstrated that significant mass transfer resistance may remain at the liquid/cellular boundary layer.

There are three major points to be stressed. First, the liquid/cellular interface may contribute significantly to mass transfer limitations. Second, when mass transfer limitations exist, the intrinsic biokinetics parameters cannot be determined. In biochemical reactor design, intrinsic parameters are essential to adequately model the system performance. Furthermore, without an understanding of the intrinsic biokinetics, one cannot accurately study transport mechanisms across biological membranes. The determination of passive or active transport across membranes is strongly affected by the extent of the liquid/cellular interfacial resistance.

1.4.2 Use in Physiologically Based Pharmacokinetics Models and Cell Culture Analog Systems

The potential toxicity of, and/or the action of, a pharmaceutical is tested primarily using animal studies. Since this technique can be problematic from both a scientific and ethical basis (Gura, 1997), alternatives have been sought. *In vitro* methods using isolated cells (Del Raso, 1993) are inexpensive, quick, and generally present no ethical issues. However, the use of isolated cell cultures does not fully represent the full range of biochemical activity as in the whole organism. Tissue slices and engineered tissues have also been studied but not without their inherent problems, such as the lack of interchange of metabolites among organs and the time-dependent exposure within the animal. An alternative to both *in vitro* and animal studies is the use of computer models based on physiologically based pharmacokinetics (PBPK) models (Connolly and Anderson, 1991). These models mimic the integrated, multicompartment nature of animals and thus can predict the time-dependent changes in blood and tissue concentrations of the parent chemical and its metabolites. The obvious limitations lie in that a response is based on assumed mechanisms; therefore, secondary and "unexpected" effects are not included. Furthermore, parameter estimation is difficult. Consequently, the need for animal surrogates or CCA systems is created. The pioneering work of M.L. Shuler's group at Cornell University (Sweeney et al., 1995; Shuler et al., 1996; Mufti and Shuler, 1998; also in Chapter 5 and the introductory comments of this book) and many others has led to the following approach.

These CCA systems are physical representations of the PBPK structure where cells or engineered tissues are used in organ compartments. The fluid medium that circulates between compartments acts as a "blood surrogate." Small-scale bioreactors housing the appropriate cell types are the physical compartments that represent organs or tissues. This concept combines attributes of PBPK and *in vitro* systems. Furthermore, it is an integrated system that can mimic dose release kinetics, conversion into specific metabolites from each organ, and the interchange of these metabolites between compartments. Since the CCA system permits dose exposure scenarios that can replicate those of animal studies, it works in conjunction with a PBPK as a tool to evaluate and modify proposed mechanisms. Thus, bioreactor design and performance evaluation testing is crucial to the success of this animal surrogate concept. Efficient transfer of substrates, nutrients, stimulants, and so on from the gas phase across all interfaces may be critical for the efficacy of certain biotransformation processes and in improving blood compatibility of biosensors monitoring the compartments. Gas/liquid mass transfer theories are well established for microbial processes (Cussler, 1984). However, biotransformation processes also involve liquid/cellular interfacial transport. In these bioreactor systems, a gaseous species is transported across two interfaces. Each could be a rate-determining step and can mask intrinsic kinetics modeling studies associated with cellular growth and/or substrate conversion and product formation.

A methanogenic chemo-autotrophic process was selected for study because of its relative simplicity and strong dependence on gaseous nutrient transport, thus establishing a firm quantitative base case (Grasso et al., 1995). The primary objective was to compare the effect of fluid hydrodynamics on mass transfer across the liquid/cellular interface of planktonic cells and the subsequent impact upon growth kinetics. Standard experimental protocol to measure the gas/liquid resistance was employed (Cussler, 1984; Grasso et al., 1995). The determination of the liquid/cellular resistance is more complex. The thickness of the boundary layer was calculated under various hydrodynamic conditions and combined with molecular diffusion and mass action kinetics to obtain the transfer resistance. Microbial growth kinetics associated with these hydrodynamic conditions can also be examined. Since Monad models are commonly applied to describe chemo-autotrophic growth kinetics (Ferry, 1993), the half-saturation constant can be an indicator of mass transfer limitations. The measured (apparent) value will be greater than the intrinsic value, as demonstrated in these earlier studies and mentioned previously.

1.5 Electro-Enzymatic Membrane Reactors as Electron Transfer Chain Biomimetics

1.5.1 Mimicry of *In Vivo* Coenzyme Regeneration Processes

In many biosynthesis processes, a coenzyme is required in combination with the base enzymes to function as high efficiency catalysts. A regeneration system is needed to repeatedly recycle the coenzyme to reduce operating costs in continuous *in vitro* synthesis processes, mimicking the *in vivo* regenerative process involving an electron transfer chain system. Multiple reaction sequences are initiated, as in metabolic cycles. NAD(H) is one such coenzyme. Because of its high cost, much effort has focused on improving the NAD(H) regeneration process (Chenault and Whitesides, 1987), with electrochemical methods receiving increased attention. The direct regeneration on an electrode has proven to be extremely difficult (Paxinos et al., 1991). Either acceleration of protonation or inhibition of intermolecular coupling of NAD$^+$ is required. Redox mediators have permitted the coupling of enzymatic and electrochemical reactions; the mediator accepts the electrons from the electrode and transfers them to the coenzyme via an enzymatic reaction, and thus regeneration/recycling of the coenzyme during a biosynthesis reaction can be accomplished (Hoogvliet et al., 1988). The immobilization of mediator and enzyme on electrodes can reduce the separation procedure, increase the selectivity, and stabilize the enzyme activity (Fry et al., 1994). Various viologen mediators and electrodes have been investigated for the NADH system in batch configurations (Kunugi et al., 1990). The mechanism and kinetics were investigated by cyclic voltammetry, rotating disk electrode, and impedance measurement techniques. The performance of electrochemical regeneration of NADH on an enzyme-immobilized electrode for the biosynthesis of lactate in a packed bed flow reactor (Fisher et al., 2000; Chen et al., 2004) is selected as a model system to illustrate an electron transfer chain biomimetic.

1.5.2 Electro-Enzymatic Production of Lactate from Pyruvate

The reaction scheme is composed of a three-reaction sequence: (1) the NADH-dependent enzymatic (lactate dehydrogenase: LDH) synthesis of lactate from pyruvate; (2) the regeneration of NADH from NAD$^+$ and enzymatic (lipoamide dehydrogenase: LipDH) reaction with the mediator (methyl viologen); and (3) the electrochemical (electrode) reaction. The methyl viologen (MV^{2+}) accepts electrons from the cathode and donates them to the NAD$^+$ via the LipDH reaction. The regenerated NADH in solution is converted to NAD$^+$ in the enzymatic (LDH) conversion of pyruvate to lactate. A key feature of this system is the *in situ* regeneration of the coenzyme NADH. A flow-by porous reactor utilizes the immobilized enzyme system (LipDH and methyl viologen as a mediator) within the porous graphite cathodes, encapsulated by a cation exchange membrane (Nafion, 124). The free-flowing fluid contains the pyruvate/lactate reaction mixture, the LDH, and the NADH/NAD$^+$ system. Lactate yields up to 70% have been obtained when the reactor system was operated in a semibatch (i.e., recirculation) mode for 24 h, as compared to only 50% when operated in a simple batch mode for 200 h. The multipass, dynamic input operating scheme permitted optimization studies to be conducted on system parameters. This includes concentrations of all components in the free solution (initial and dynamic input values could be readily adjusted through recycle conditioning), flow rates, and electrode composition and their transport characteristics. By varying the flow rates through this membrane reactor system, operating regimes can be identified that determine the controlling mechanism for process synthesis (i.e., mass transfer vs. kinetics limitations). Procedures for operational map development are thus established.

References

Alberts, B., D. Bray, J. Lewis, M. Raff, K. Roberts, and J. D. Watson. 1989. *Molecular Biology of the Cell*, 2nd ed., Garland, New York.

Chen, X., J.M. Fenton, R.J. Fisher, and R.A. Peattie. 2004. Evaluation of in-situ electro-enzymatic regeneration of co-enzyme NADH in packed bed membrane reactors: Biosynthesis of lactate, *J. El. Chem. Soc.*,151(2):236–242.

Chenault, H. K. and G. M. Whitesides. 1987. Regeneration of nicotinamide cofactors for use in organic synthesis, *Appl. Biochem. Biotechnol.* 14:147–197.

Connolly, R.B. and M.E. Anderson. 1991. Biologically based pharmacodynamic models: Tool for toxicological research and risk assessment, *Annu. Rev. Pharmacol. Toxicol.*, 31:503.

Cussler, E.L. 1984. *Diffusion: Mass Transfer in Fluid Systems*, Cambridge University Press, New York.

Cussler, E., R. Aris, and A. Bhown. 1989. On the limits of facilitated diffusion, *J. Membr. Sci.*, 43:149–164.

Del Raso, N.J. 1993. *In vitro* methodologies for enhanced toxicity testing, *Toxicol. Lett.*, 68:91.

Dinno, M.A., R.J. Fisher, J.C. Matthews, L.A. Crum, and W. Kennedy. 1991. Effects of ultrasound on membrane bound ATPase activity, *J. Acous. Soc. Am.* 90(4):2258–2262.

Ferry, J.G. 1993. *Methanogenesis*, Chapman and Hall, New York.

Fisher, R.J. 2000. Compartmental analysis, Chapter 8 in *Introduction to Biomedical Engineering*, (Editors) J. Enderle, S. Blanchard and J. Bronzino, Academic Press, Orlando, FL.

Fisher, R.J., J.M. Fenton, and J. Iranmahboob 2000. Electro-enzymatic synthesis of lactate using electron transfer chain biomimetic membranes, *J. Membr. Sci.*, 177:17–24.

Freshney, R.I. 2000. *Culture of Animal Cells*, 4th Edition, Wiley-Liss, New York.

Fry, A.J., S.B. Sobolov, M.D. Leonida, and K.I. Viovodov. 1994. Immobilization of mediator and enzymes on electrodes for NADH regeneration, *Denki Kagaku*, 62:1260.

Galletti, P.M., C.K. Colton, M. Jaffrin, and G. Reach 2000. Artificial pancreas, Chapter 134 in *Bio Medical Engineering Handbook*, 2nd Ed, (Editor) J. Bronzino, CRC Press, Boca Raton, FL.

Giusti, P., L. Lazzeri, and L. Lelli. 1993. Bioartificial polymeric materials, *TRIP*, 1(9):21–25.

Grasso, K., K. Strevett, and R. Fisher. 1995. Uncoupling mass transfer limitations of gaseous substrates in microbial systems, *Chem. Eng. J.*, 59:195–204.

Gura, T. 1997. Systems for identifying new drugs are faulty, *Science*, 273:1041.

Hoogvliet, J.C., L.C. Lievense, C. V. Kijk, and C. Veeger. 1988. Regeneration of co-enzymes in biosynthesis reactions, *Eur. J. Biochem.*, 174:273.

Jain, R.K. 1994. *Transport Phenomena in Tumors, Advances in Chemical Engineering*, Vol. 19, pp. 129–194, Academic Press, Inc., San Diego.

Johnson, A.E., R.J. Fisher, G.C. Weir, and C.K. Colton. 2009. Oxygen consumption and diffusion in assemblages of respiring spheres: Performance enhancement of a bioartificial pancreas, *Chem. Eng. Sci.*, 64(22):4470–4487.

Kalachev, A.A., L.M. Kardivarenko, N.A. Plate, and V.V. Bargreev. 1992. Facilitated diffusion in immobilized liquid membranes: Experimental verification of the jumping mechanism and percolation threshold in membrane transport, *J. Membr. Sci.*, 75:1–5.

Kumar, R., R. Tyagi, V.S. Parmar, A.C. Watterson, J. Kumar, J. Zhou, M. Hardiman, R. Fisher, and C.K. Colton. 2005. Perfluorinated amphiphilic polymers as nano probes for imaging and delivery of therapeutics for Cancer, *Polym Prepr.*, 45(2):230–233.

Kunugi, S., K. Ikeda, T. Nakashima, and H. Yamada. 1990. Enzyme electrode based on gold-plated polyester cloth, *Poly Bull*, 24:247.

Lauffenburger, D.A. and J.J. Linderman. 1993. *Receptors: Models for Binding, Trafficking, and Signaling.* Oxford University Press, New York.

Lewis, A.S. and C.K. Colton. 2004, Tissue engineering for insulin replacement in diabetes, in *Scaffolding in Tissue Engineering*, (Editors) Ma, P.X. and Elisseeff, J., Marcel Dekker, New York.

Luo, Y. and G.D. Prestwich. 2001. Hyaluronic acid-*N*-hydroxysuccinimide: A useful intermediate for bio-conjugation, *Bioconjug. Chem.*, 12:1085–1088.

Michaels, A.L. 1988. Membranes, membrane processes and their applications: Needs, unsolved problems, and challenges of the 1990s, *Desalination*, 77:5–34.

Matthews, J.C., W.L. Harder, W.K. Richardson, R.J. Fisher, A.M. Al-Karmi, L.A. Crum, and M.A. Dinno. 1993. Inactivation of firefly luciferase and rat erythrocyte ATPase by ultrasound, *Membr. Biochem.*, 10:213–220.

Mufti, N.A. and M.L. Shuler. 1998. Different *in vitro* systems affect CYPIA1 activity in response to 2,3,78-tetrachlorodibenzo-*p*-dioxin, *Toxicol. In Vitro*, 12:259.

Palsson, B. 2000. Tissue engineering, Chapter 12 in *Introduction to Biomedical Engineering*, (Editors). J. Enderle, S. Blanchard, and J. Bronzino, Academic Press, Orlando, FL.

Panagiotou, T. and R.J. Fisher. (2008 October). Form nanoparticles via controlled crystallization: A "bottom-up" approach, *Chem. Eng. Prog.*, 10:33–39.

Panagiotou, T., S. Mesite, and R.J. Fisher. 2009. Production of norfloxacin nano-suspensions using Microfluidics Reaction Technology (MRT) through solvent/anti solvent crystallization, *Ind. Eng. Chem. Res.*, 48(4):1761–1771.

Paxinos, A.S., H. Gunther, D.J.M. Schmedding, and H. Simon. 1991. Direct electron transfer from modified glassy carbon electrodes carrying covalently immobilised mediators to a dissolved viologen accepting pyridine nucleotide oxidoreductase and dihydrolipoamide dehydrogenase, *Bioelectro. Bioenerg.*, 25:425–436.

Peattie, R.A., A.P. Nayate, M.A. Firpo, J. Shelby, R.J. Fisher, and G.D. Prestwich. 2004. Stimulation of *in vivo* angiogenesis by cytokine-loaded hyaluronic acid hydrogel implants, *Biomaterials*, 25:2789–2798.

Rabinow, B. 2004. Nanosuspensions in drug delivery, *Nat Rev-Drug Discov.*, 3:785–796.

Rabinow, B. 2005. Pharmacokinetics of drugs administered in nanosuspensions, *Discov Med.*, 5(25):74–79.

Schwarzer, H.C. and W. Peukert. 2004. Tailoring particle size through nanoparticle precipitation, *Chem. Eng. Comm.*, 191:580–608.

Shuler, M.L., A. Ghanem, D. Quick, M.C. Wang, and P. Miller. 1996. A self-regulating cell culture analog device to mimic animal and human toxicological responses, *Biotechnol. Bioeng.*, 52:45.

Srinivasan, A.V., G.K. Haritos, and F.L. Hedberg. 1991. Biomimetics: Advancing man-made materials through guidance from nature, *Appl. Mechanics Reviews*, 44:463–482.

Stockton, E. and R.J. Fisher. 1998. Designing biomimetic membranes for ion transport, *Proc. NEBC/IEEE Trans.*, 24(4):49–51.

Sweeney, L.M., M.L. Shuler, J.G. Babish, and A. Ghanem. 1995. A cell culture analog of rodent physiology: Application to naphthalene toxicology, *Toxicol. In Vitro*, 9:307.

Thoresen, K. and R.J. Fisher. 1995. Use of supported liquid membranes as biomimetics of active transport processes, *Biomimetics*, 3(1):31–66.

Weiss, R.A., C. Beretta, S. Sasonga, and A. Garton. 1995. Applications for ionomers, *Appl. Polym. Sci.*, 41:491.

2

Transport/Reaction Processes in Biology and Medicine

2.1	Introduction	2-1
2.2	Macroscopic Approximations: Allometry................................	2-2
2.3	Orders of Magnitude and Characteristic Time Constants	2-3
2.4	Time Constant Ratios	2-5
	Classification of Blood Vessels • Simultaneous Diffusion and Chemical Reaction	
2.5	Systems of Multiple Time Constants...................................	2-7
	Importance of Boundary Conditions • Pharmacokinetics and Related Processes • Hemodialysis • Gene Expression in Prokaryotes • Exercise and Type II Diabetes	
2.6	Pseudocontinuum Models	2-15
	Tissue Oxygenation • Pulmonary Structure and Function • Pulmonary Blood-Gas Matching	
2.7	More Complex Situations...	2-17
	Stochastic Behavior of Genetic Regulation • Cellular Crowding • The Dual Nature of Oxygen • Self-Organization and Emergence	
References..		2-18

E. N. Lightfoot
University of Wisconsin

2.1 Introduction

Transport phenomena, and particularly mass transfer and chemical reaction, govern a great variety of physiological and pathological processes, and they supplement in a nontrivial way genetic factors in both organism development and species evolution. For humans and other mammals, of primary interest here, the body may in fact be viewed as a complex and hierarchical transport/reaction system supplying the needs of genes and protecting them from the external environment. This is suggested in Figure 2.1 where four major organs are seen to interact directly with the external environment and, via an extremely complex series of mass transport processes, with other organs, all body cells, and ultimately their genes. These processes range from convective transport via blood and pulmonary gases to extremely complex forced diffusion mechanisms at the cellular and subcellular level.

Our purpose here is to suggest effective bases for modeling and manipulating selected subsystems of living bodies, and it must be recognized at the outset that a complete description is impossible. Just a glance at an atlas of the human anatomy will make this clear. The body will always be what Malcolm Gladwell [26] calls a mystery: a problem for which no complete solution exists. Our approach is to suggest approximations simple enough to be soluble, testable, and hopefully of some utility: what Gladwell calls

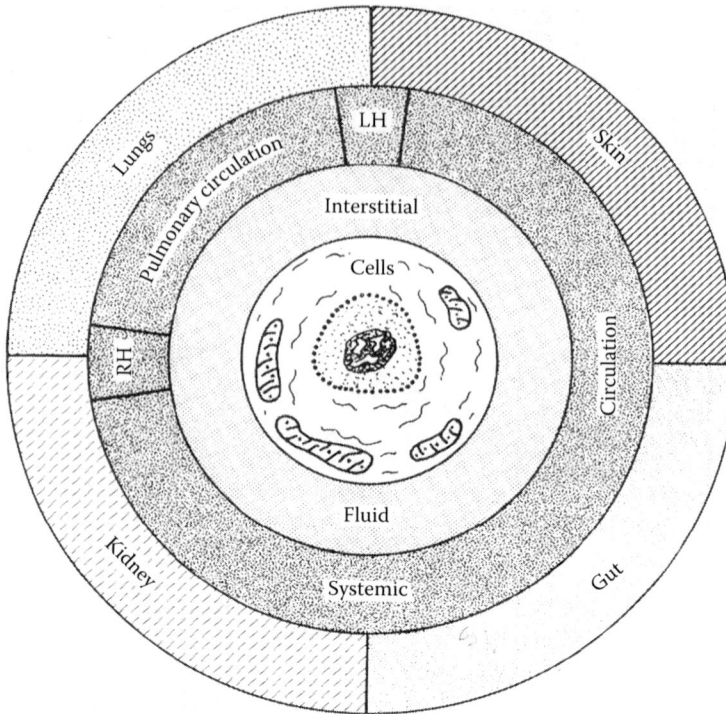

FIGURE 2.1 Mammalian topology.

puzzles. A major review of multiscale modeling in biological systems is provided by Bassingthwaighte et al. [3], and we must settle for a much shorter discussion here. We start with macroscopically available parameters and proceed to more detailed descriptions.

2.2 Macroscopic Approximations: Allometry

We begin with observable relationships between macroscopic quantities, and express these by equations of the *allometric* form [14]:

$$P = aM^b$$

Here, "P" is some property to be predicted, "M" is a known property, and "a" and "b" are constants either already known or to be determined. Most commonly, M is the mass of the body or organ under consideration. The correlation of basal metabolic rates with body mass shown in Figure 2.2 is perhaps the most commonly available example, and it may be seen that even here data scatter is large. Under basal conditions, fat is the primary body fuel, and heat generation is about 4.7 kcal/mmol oxygen (STP). It is suggested that only recent correlations be used, for example, those of White and Seymour [62] or Roberts et al. [44]. Variants abound, and many are unreliable. Many mechanisms have been suggested as responsible for this correlation, but it should really be regarded as empirical. However, a few remarks are in order.

A primary reason that *basal metabolic rates* increase slower than linearly with body mass is that the proportion of highly active organs, for example, brain, liver, and kidney, falls with increase in total mass. Thus, total brain mass for most mammals increases more slowly than body mass [14,50, Table 3.4, p. 48]:

$$M_{\text{Brain}} = 0.011M^{0.76}$$

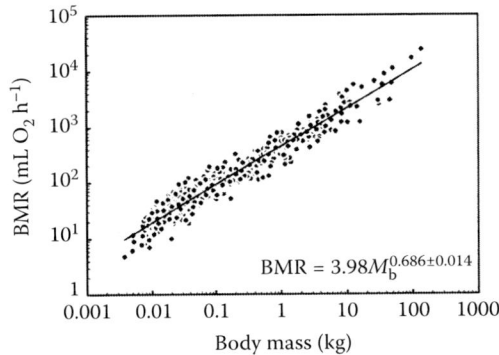

FIGURE 2.2 Basal metabolic rate.

TABLE 2.1 Brain Allometry

Species	Mammals	Monkeys	Great Apes	Humans
Coefficient (a)	0.01	0.02–0.03	0.03–0.04	0.08–0.09
Exponent (b)	0.7	0.66	0.66	0.66

Here, both masses are in kilograms. However, *overall* brain oxygen consumption per unit mass is invariant at about [39–41]

$$R_{O_2}(\text{brain}) = 3.72 \times 10^{-5}\,\text{mmols O}_2/\text{mL,s}$$

The activity of liver and kidney, as well as their proportionate mass, do fall off with increasing size, but this is at least in part because more ducts and mechanical support are needed for larger animals.

However, primate brains are much bigger than those for other mammals, and ours are bigger yet (see Table 2.1 [49]).

It may be seen from comparing the table entry for mammals with the above equation that allometric correlations are far from exact. The real message of this table is that species differences can be very large, and that our major asset relative to other animals is mental. All species deviate from the mean in some respect, and this must be expected.

For most species maximum metabolic rate is about 10 times basal [33, p. 212]. However, some very athletic animals have much higher ratios of maximal to basal rates: pronghorn antelopes can achieve ratios of 65, and alligators can reach 40. A very informative discussion of diet, metabolic rates, and temperature control is provided in Chapter 8 of Reference 32.

It is also extremely important to determine in so far as possible the conditions under which the data presented have been determined and for which they will be used. A recently identified example of major uncertainty is that of gender differences between male and female rats and mice [56]. Planning and interpreting animal experiments are primary applications of allometry, but they must be used with care.

It is also clear that we must take a deeper look, and we begin immediately below.

2.3 Orders of Magnitude and Characteristic Time Constants

During almost all serious creative endeavors, the potential parameter space of interest is unmanageably large, and it is often desirable to limit initial consideration to *orders of magnitude* of probable relevance. Even the term "order of magnitude" is unfortunately heuristic, and success here depends in the last analysis on experience and judgment of the investigators. We try here to provide helpful examples based

TABLE 2.2 Commonly Used Diffusion Lengths

Sphere	Radius R	$\ell^2 = R^2/6$	>0.99 complete
Cylinder	Radius R	$R^2/4$	>0.99
Slab	Half thickness L	$L^2/2$	>0.93

upon our own experience. Time constants prove to be of particular importance, and we use these as the primary basis for examples here.

We shall moreover consider just three types of time constants: mean residence times, t_m, diffusional times, $\ell^2/D_{im} = t_{dif}$, and reaction times, $c_i/R_i = t_{rxn}$. Here, "t" is time, "ℓ" is a characteristic length, "D_{im}" is a characteristic diffusivity of species "i" through mixture "m," "c_i" is local molar concentration of species "i," and R_i is the local molar rate of production of "i" by chemical reaction.

Mean residence time is only well defined for a volume V with impermeable surfaces except for one inlet and one outlet, with time-independent volumetric flow rate Q through it, and negligible diffusion across the inlet and outlet. Under these circumstances, mean residence time is given by [48,11, p. 756] (see Table 2.2)

$$t_m = V/Q$$

Characteristic length squared for the diffusional time constants are usually adapted to system shape.

Degrees of completion are calculated for uniform initial concentration and zero concentrations on the object boundary surface. Many other situations are described in such references as the venerable but still useful Reference 17.

As an example, we consider transients in the alveoli of the human lung: irregular sacs at the distal ends of the branching pulmonary system. To ensure a conservative estimate, we shall treat them as flat plates of half thickness 0.0105 cm, and we note that effective diffusivity of oxygen in pulmonary air is close to 0.2 cm²/s. The diffusional response time here is then

$$t_{dif} = (0.015\,\text{cm})^2/(0.2 \times 2\,\text{cm}^2/\text{s}) = 0.56\,\text{ms}$$

This is very short compared to the 1 s residence time of alveolar blood and the 1/12 of a minute between breaths. We may assume the alveolar gas to be well mixed.

It must be kept in mind that intracellular properties may differ substantially from those in saline solutions of the same ionic strength. This is in large part because cell interiors are crowded and contain membranes that limit mass transport. They also contain structures such as mitochondria that are impermeable to many diffusing metabolites. A few representative examples are supplied in Table 2.3.

TABLE 2.3 Intracellular Diffusion Coefficients

Compound	MW	Radius AA	Diffusivity (10^{-7} cm²/s) Water	Cells	Ratio Water/Cell
Sorbitol	182	2.5	94	50	1.9
Methylene blue	320	3.7	40	15	2.6
Sucrose	324	4.4	52	20	2.6
Eosin	648	6	40	8	5
Dextran	3600	12	18	3.5	5
Inulin	5500	13	15	3	5
Dextran	10,000	23.3	9.2	2.5	3.7
Dextran	24,000	35.5	6.3	1.5	4.2
Actin	43,000	23.2	5.3	0.03	167
Bovine serum albumin	68,000	36	6.9	0.1	65–71

Source: Adapted from Mastro, A. M. et al. 1984. *PNAS*, 81, 3414–3418.

Most of the systems of interest to us involve more than one time constant, and that is even true of our simple example of alveolar transients: these are really only of interest by way of comparison with blood circulation and breathing rates.

2.4 Time Constant Ratios

Many biological transport processes are primarily determined by ratios of time constants, and it is useful to look at them in this light. Moreover, examination of graphs in such texts as *Transport Phenomena* shows that the abscissas of a great many are themselves ratios of time constants, and recognizing this can often provide useful insight. Very often, for example, the behavior of both physiological and well-designed artificial systems tends to cluster about abscissa magnitudes of the order of unity. We now look at some specific examples.

2.4.1 Classification of Blood Vessels

We base our analysis on the canine data of Table 2.4 [16 and since reproduced in many newer sources, e.g., 25]. These are reasonably representative of their human counterpart except in lacking the larger arteries. The latter are of no interest to the present discussion as we are here concerned only in transport of respiratory gases and other small solutes between blood and surrounding tissue. Note that a Schmidt number of 1000 is used in this table. This is reasonable at the order-of-magnitude level of discussion.

We can extrapolate the information in this table by recognizing that most arteries end in binary branches, and to a surprisingly good approximation, they follow Murray's law:

$$R_2^3 = R_1^3/2$$

Here, R_2 is the radius of the two daughter arteries and R_1 that of the mother [52, p. 61], and it is shown by Fung [25, p. 118] that this tends to minimize the cost of blood flow: arteries serve primarily as transporters for blood and many other metabolites, and both construction and operation use a great deal of metabolic energy. Moreover, their diameters are continually controlled to maintain a desired wall shear stress ([25, p. 494] and [52, p. 67]). Veins parallel arteries and are larger in diameter. They provide storage as well as transport, and their diameters can be adjusted in responses to changes in blood volume.

We limit discussion to the interplay of flow and transverse mass transport, and we now estimate the utility of the various vessels to transmit dissolved substances to their walls by diffusion. We begin by looking at the last column in the table.

TABLE 2.4 Classification of Blood Vessels

Vessel	D (cm)	L	$<v>$ Mean	L/D	Re Mean	Sc Assumed	L/D ReSc
Ascending aorta	1.5	5	20	3.333333	4500	1000	7.40741E−07
Descending aorta	1.3	20	20	15.38462	3400	1000	4.52489E−06
Adominal aorta	0.9	15	15	16.66667	1250	1000	1.33333E−05
Femoral artery	0.4	10	10	25	1000	1000	0.000025
Carotid artery	0.5	15		30		1000	
Arteriole	0.005	0.15	0.75	30	0.09	1000	0.333333333
Capillary	0.0006	0.06	0.07	100	0.001	1000	100
Venule	0.004	0.15	0.35	37.5	0.035	1000	1.071428571
Inferior vena cava	1	30	25	30	700	1000	4.28571E−05
Main pulmonary artery	1.7	3.5	70	2.058824	3000	1000	6.86275E−07

Source: Abstracted from Fung, Y. C. 1997. *Biomechanics: Circulation*, Springer, New York, Table 3.1:1, p.110.

$$L/D\,\mathrm{Re}\,\mathrm{Sc} = (L/D)(\mu/D <v> \rho)(D_{im}\rho/\mu)$$
$$= (L/<v>)/(D^2/D_{im})$$

Here, $<v>$ is the mean or flow average velocity [11, 2.3–20, p. 51]. Now

$$L/<v> = t_m; \quad D^2/D_{im} = 16 t_{diff}$$

We may thus write

$$t_m/t_{dif} = 16 L/D\,\mathrm{Re}\,\mathrm{Sc}$$

for small solutes such as gases.

It is now clear that only capillaries, and to a small degree arterioles and venules, are capable of transferring solutes between themselves and surrounding tissue. These three classes have long been known as the microcirculation because they are invisible to the naked eye. We now see a functional basis for this classification.

2.4.2 Simultaneous Diffusion and Chemical Reaction

We now base our discussion on a variant of the familiar effectiveness chart with an abscissa containing only observable quantities [59], Figure 2.3. Here, effectiveness factors are shown as a function of a Weisz modulus

$$\Phi = \frac{\langle R_i \rangle (V/S)^2}{c_i D_{im}}$$

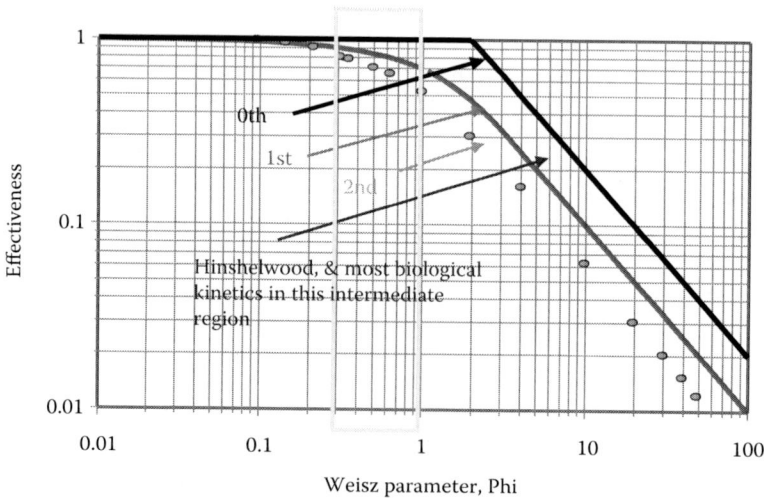

$$\Phi = \frac{<R_i> (V/S)^2}{c_{i0} \mathcal{D}_{im}} \equiv \frac{T_{dif}}{T_{rxn}}$$

FIGURE 2.3   Effectiveness factors.

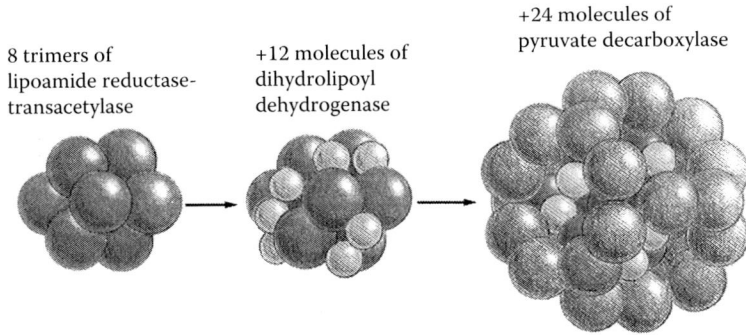

8 trimers of
lipoamide reductase-
transacetylase

+12 molecules of
dihydrolipoyl
dehydrogenase

+24 molecules of
pyruvate decarboxylase

FIGURE 2.4 **(See color insert.)** The pyruvate dehydrogenase complex.

for a porous catalyst in the form of a slab. Here, $<R_i>$ is the observed rate of reaction of species "i" per unit volume, $<S/V>$ is the observable specific surface of the catalyst (half thickness for a slab), c_i is the concentration of species "i" in the feed to the catalyst, and D_{im} is the effective diffusivity of species "i" through the catalyst matrix. Lines are shown for zeroth and first-order irreversible reaction and dots for second order. Michaelis–Menten and most biological kinetics are represented in the small region between zero and first order. To our present order of approximation, this graph is also valid for other shapes: $V/S = R/3$ for spheres; $R/2$ for cylinders.

It was shown by Weisz [59] that biological systems, and also well-designed industrial catalysts, tend to exhibit Weisz parameters between 1/3 and 1. However, as suggested in Figure 2.4 [1], biological systems tend to be much more complex. Putting catalysts for successive reactions adjacent to one another as in Figure 2.4 can make up for thermodynamically unfavorable intermediates, as suggested, for example, by Weisz [59], and nesting them decreases the chance for escape. The advantages of complex parallel reactions are discussed by Lane [32, p. 26] in connection with the Krebs cycle.

2.5 Systems of Multiple Time Constants

The mean residence time of solute in a flow system is of course only a partial description and corresponds to the first temporal moment [11, 23.6–3, p. 756] of exit tracer concentration for a pulse tracer input. It says nothing about the distribution of exit concentration as a function of time or the effect of the internal flow behavior. However, there are many circumstances where the shape is of no importance, and we begin by identifying some of these.

2.5.1 Importance of Boundary Conditions

We show an example of shape insensitivity by comparing the behavior of the two commonly used models shown in Figure 2.5: continuous stirred tank reactors, CSTRs, and plug flow reactors, PFRs. Here, the mean residence time of the test system is again defined as t_m.

The upper graph in this figure shows that the shapes of the exit tracer distributions for a pulse input are very different for these two systems. However, the responses to exponentially decaying input tracer concentrations

$$c_{in}(t) = c_0 e^{-t/t_0}$$

are much more interesting. Here, t_0 is the decay constant for the input stream. If

$$t_m = t_0$$

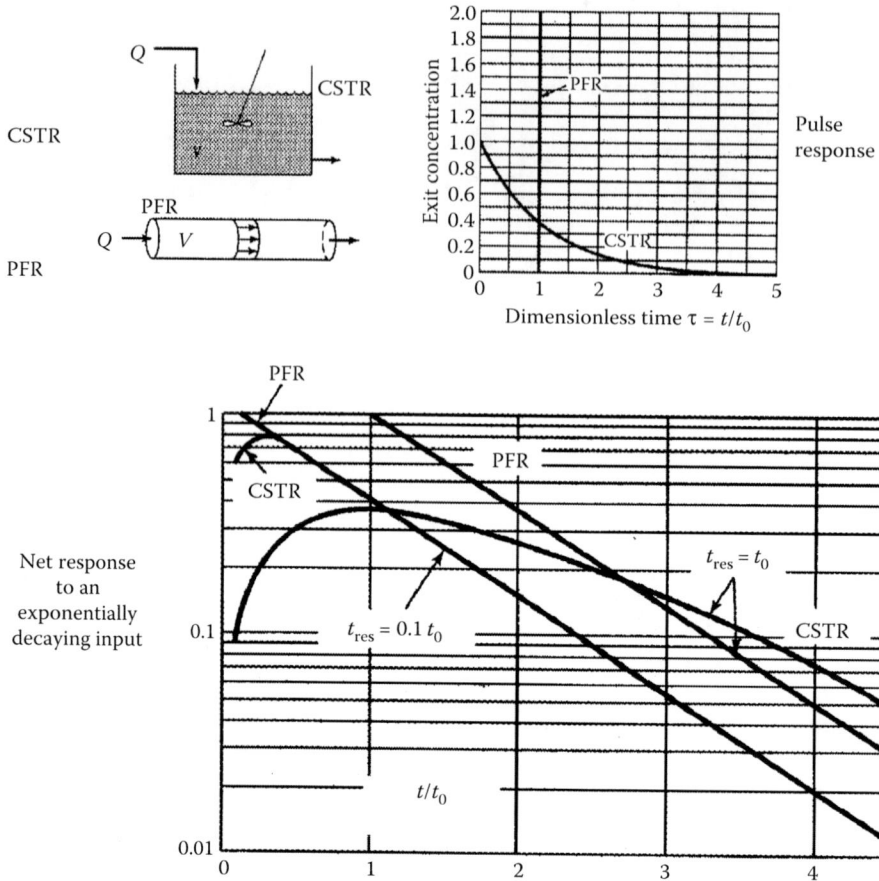

FIGURE 2.5   Universality.

the responses of the two flow systems are still quite different. However, if

$$t_m \ll t_0$$

the PFR and CSTR responses are indistinguishable, except at very short observer times. Moreover, they are simple decaying exponentials.

This simplification results from *time constant separation*, and it is characteristic of physiological, as opposed to pathological, systems. It is shown, for example, in glycolysis [30], Figure 2.6. There are 10 reactions shown here, and seven of them are so rapid that they may be considered instantaneous. System behavior is then controlled by three reactions, labeled HK, PFK, and PK, which are much slower.

2.5.2 Pharmacokinetics and Related Processes

The time scale separation characteristic of physiological systems is responsible for the success of many macroscopic descriptions and notably pharmacokinetics: that branch of pharmacology dealing with the absorption, distribution, metabolism, and elimination of drugs and toxins [43, 60 with others constantly appearing]. The body is now approximated by a network of mixing tanks as suggested in Figure 2.7. Here, each box shown is assumed to be in complete equilibrium, internally and with the venous blood leaving it. Models of this general type had long been used by anesthesiologists, the diving community,

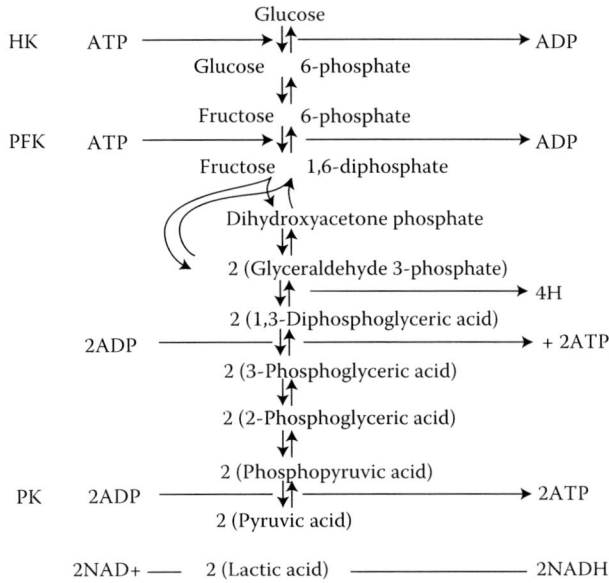

FIGURE 2.6 Glycolysis.

and others, but each box was completely empirical and characterized only by an assumed mean residence time t_m. The residence times were chosen empirically to fit observation.

A major step forward was made by two chemical engineers, the late Ken Bischoff [12] and Bob Dedrick [13], who identified the boxes as organs or groups of organs and predicted their mean residence times from actual physiological data. It was now possible to make *a priori* predictions and to

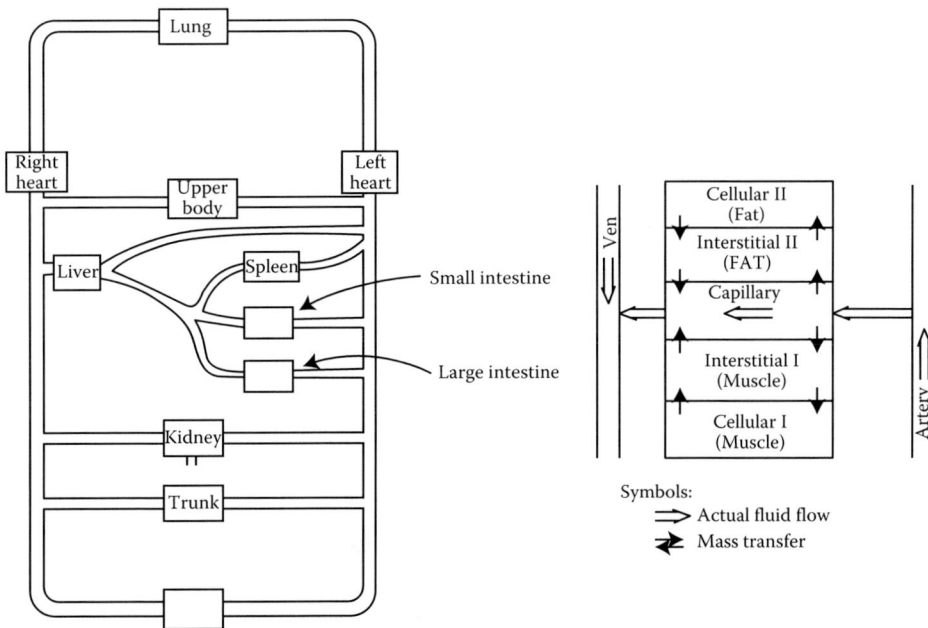

FIGURE 2.7 Pharmacokinetic approximations.

explain the physiology upon which the resulting models were based. It was even possible to allow for concentration dependence of residence times and thus to describe the behavior of nonlinear systems with confidence.

Unfortunately, we shall have space here only for two simple examples, but the texts suggested above and others that may become available can extend this discussion enormously. An important related area is that of whole organ models [4]. See also References 29,55.

2.5.3 Hemodialysis

We first turn our attention to the amelioration of end-stage kidney disease by dialysis of blood against electrolyte solutions containing a healthy balance of metabolites [11, p. 733]. The nephrons that are to be supplemented and a sketch of a typical dialyzer are shown in Figure 2.8. The flat plate dialyzer shown is normally replaced by bundles of hollow fibers, and sometimes by the patient's peritoneal membrane. The dialyzing solution is normally prepared at the patient's side by dilution of a concentrate with carefully purified water, and accurate temperature control is needed for control as the dialysate composition is monitored by conductivity.

The system model is shown in Figure 2.9, and the conservation equations for the two compartments are

$$V_T dc_T/dt = G - Q(c_T - c_B)$$
$$V dc_B/dt = Q(c_T - c_B) - Clc_B$$

Here, V_T is the total amount of tissue fluid, excluding that of the blood, V_B is that in the blood, c_T is the concentration of the metabolite of interest in the tissue, and c_B is that in the blood, G is the total rate of metabolite formation (assumed constant), and Cl is the "clearance of the dialyzer": the fraction of

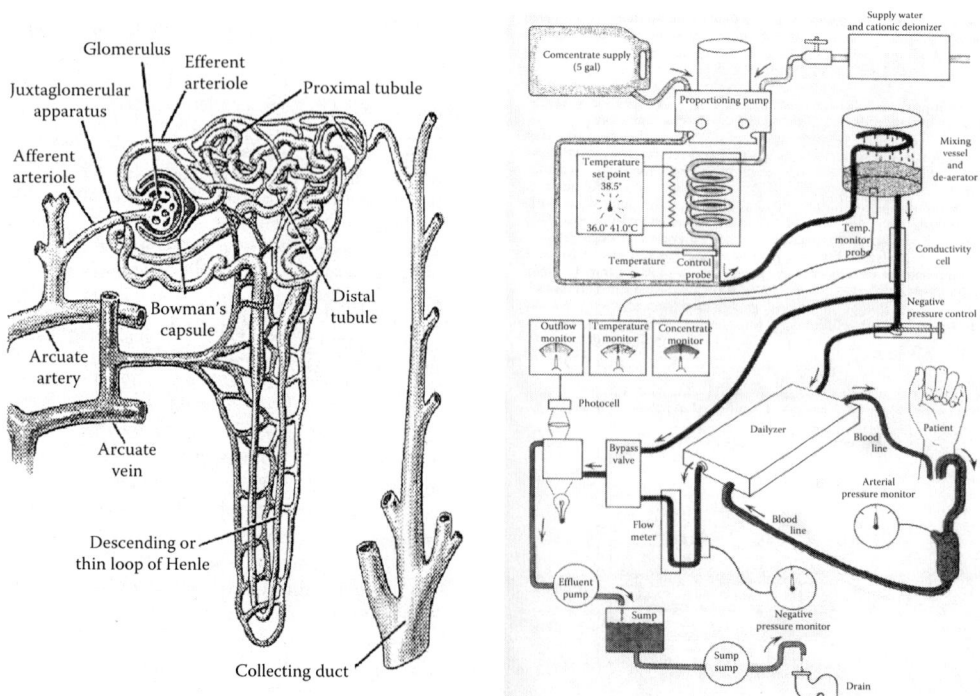

FIGURE 2.8   Hemodialysis.

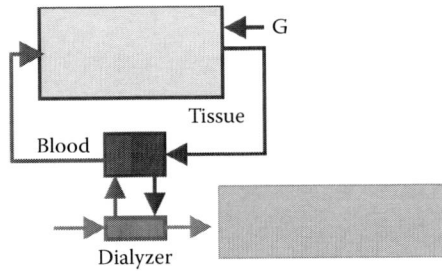

FIGURE 2.9 Dialysis model.

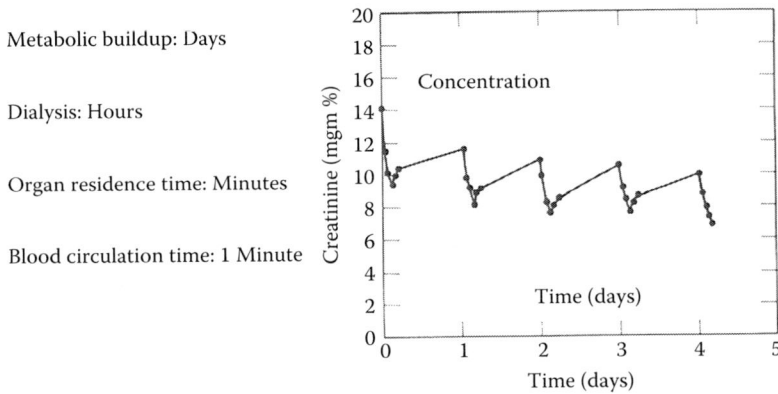

Metabolic buildup: Days

Dialysis: Hours

Organ residence time: Minutes

Blood circulation time: 1 Minute

FIGURE 2.10 Test of creatinine modeling.

metabolite removed in one pass. Since the dialysate enters metabolite-free, clearance may be considered a property of the dialyzer and treated as a constant. Initially, the metabolite concentrations in blood and tissue may be considered equal.

Shown in Figure 2.10 are data for creatinine in a test dialyzer. The needed parameters are obtained to fit data for the first of the five dialyses shown. It may be seen from this and the four subsequent dialyses that the model fit is excellent. Clearly, the model is greatly simplified: neither the body tissue nor the blood is a simple mixing tank. Moreover, the rate of metabolite formation is far from constant. Rather the goodness of fit is due to excellent time constant separation: metabolite formation (day), dialysis (hour), organ residence time (minutes), and blood circulation (1 min).

Generally, time scales more than about a factor three or less than about one-third relative to that of primary interest can be assumed never to take place or to be instantaneous, respectively. Note that

$$e^3 = 20.1$$

These "rules of thumb" for order-of-magnitude analysis are clearly met in this example. However, it is always wise to check such assumptions to the extent possible and the creatinine data do this very nicely here.

2.5.4 Gene Expression in Prokaryotes

We next consider the dynamics of gene expression in bacterial cells as suggested in Figure 2.11. Here, a regulator, a protein molecule, must diffuse from an arbitrary initial position to activate a gene, that may

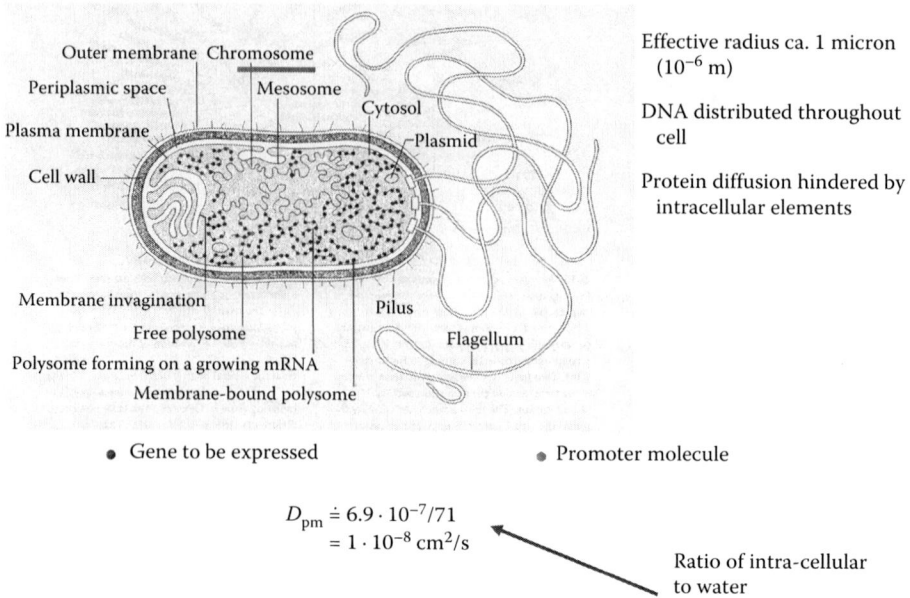

Outer membrane   Chromosome
Periplasmic space   Mesosome
Plasma membrane   Cytosol
                                       Plasmid
Cell wall

Membrane invagination   Pilus
Free polysome   Flagellum
Polysome forming on a growing mRNA
Membrane-bound polysome

Effective radius ca. 1 micron (10^{-6} m)

DNA distributed throughout cell

Protein diffusion hindered by intracellular elements

● Gene to be expressed   ● Promoter molecule

$$D_{pm} \doteq 6.9 \cdot 10^{-7}/71$$
$$= 1 \cdot 10^{-8} \text{ cm}^2/\text{s}$$

Ratio of intra-cellular to water

FIGURE 2.11 Gene expression in a bacterial cell.

also be anywhere in the cell [9,37]. We assume that this operation is a diffusion-controlled reaction that can be described by three sequential steps:

1. Sampling the entire region of the cell. For our present purposes, the regulator, the active portion of the gene, and the cell will all be treated as spherical. The radii of both the regulator and gene are taken as 2.5 nm and that of the cell as 1 micron. These simplifications are only reasonable at the order-of-magnitude level, but we shall find them to suffice. The time required for this process is of the order

$$T_{cell} \sim R_{cell}^2/6D_{PC}$$

Now, the diffusivity of the regulator through the cytoplasm will be far lower than in saline solution. We take it here to be that of serum albumin (see Table 2.3):

$$D_{PC} = 10^{-8} \text{ cm}^2/\text{s}; \quad T_{cell} \sim 10^{-8} \text{ cm}^2/6 \times 10^{-8} \text{ cm}^2\text{s}^{-1} = (s/6)$$

This is very fast.

2,3. Diffusion to the gene, now treated as a sphere of the combined radii of 5 nm, across a boundary layer. Here, we will use the expression for mass flux of the regulator protein reaching the gene from large surroundings across this surface (author):

$$N_P = \frac{D_{PC}c_{P\infty}}{(R_P + R_G)}\left[1 + 1/\sqrt{\pi\tau}\right]; \quad \tau = tD_{PC}/(R_P + R_G)^2$$

We first note that the transient term

$$1/\sqrt{\pi 10^{10} t/s}$$

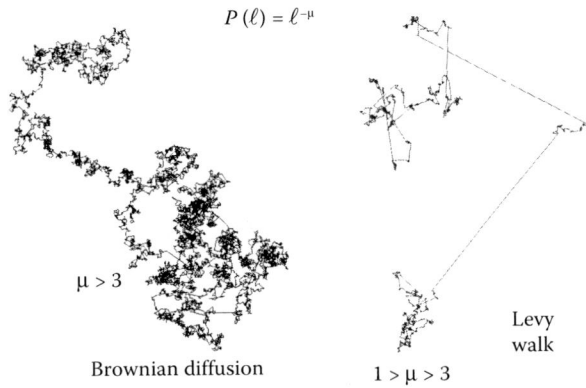

FIGURE 2.12 Dispersion models.

may be safely ignored. Now the probability P of finding the regulator still diffusing freely in the cytoplasm is given by a mass balance:

$$V_{cell} \frac{dc_P}{dt} = A_{sph} N_P; \quad d\ln cP/dt = 3 D_{PC} \left[(R_P + R_G)/R_{cell}^3 \right] = 1/T_{rxn}$$

Then

$$T_{rxn} = \left[\frac{1.5 \times 10^{-14}}{(10^{-4})^3} \right]^{-1} \text{s} = 67\,\text{s}$$

This is by far the longest of the three characteristic times, and it governs the reaction. The above order-of-magnitude-based procedure is far simpler than the numerical approach used by the initial investigators [9], and it leads to the same result.

However, gene expression was found experimentally to be much faster than predicted. The authors suggested this speed resulted from weak adsorption of the regulator on all sections of the gene. As a result it could move in a one-dimensional dispersion along the gene on contact, then release itself, and sample other regions. This explanation seems to have been accepted and not examined further. We see then that dispersion in biological systems can be much more complicated than has been observed in more conventional situations.

One major example that has been substantiated is foraging, for example, for food supplies that are not uniformly distributed [53]. An example shown in Figure 2.12 contrasts Brownian dispersion with a Levy walk. It has been found that a great many species, from large animals to microscopic organisms, use something very close to a Levy walk [10,36,53] with a "mu" parameter close to the optimum magnitude of 2. Here, the probability P of a "jump" of length "ℓ" is shown as

$$P(\ell) = \ell^{-\mu}$$

Application of Levy walk theory in biology is still new, and medical applications are not yet known to this author. However, it would be surprising if it did not prove useful in studying epidemics and other dynamic disease situations.

2.5.5 Exercise and Type II Diabetes

Time constants can be useful even for nonlinear systems, but analysis here must be simplified. We use diabetes control by patients here as an important system, and one too complex for detailed modeling. A comparison of glucose removal from the blood stream between diabetic and nondiabetic ("normal") patients is given in Figure 2.13.

We start by noting that blood glucose levels following meals for diabetics reach a plateau an hour or two after a meal and then decrease very slowly. This decrease can be greatly accelerated, and extended, by several minutes of moderate aerobic exercise as suggested in Figure 2.14 (data of author). Muscle contraction acts much like insulin in moving the glucose transporters, "GLUT4," to the surface of muscle calls [36,38,46,47, and many newer]. A practical test of this effect is shown in Figure 2.15 (data of author). Here, 20–30 min brisk walks were taken about 2 h after breakfast in addition to normal doses

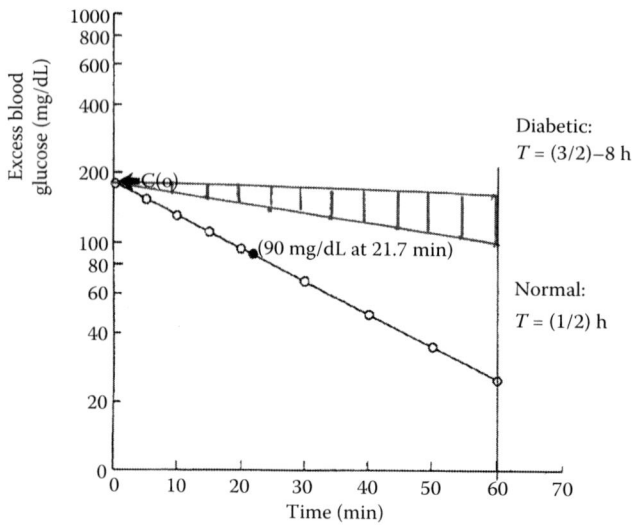

FIGURE 2.13   Glucose tolerance test.

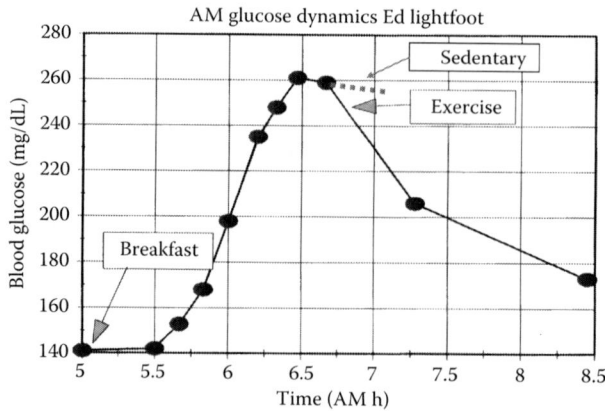

FIGURE 2.14   Glucose dynamics.

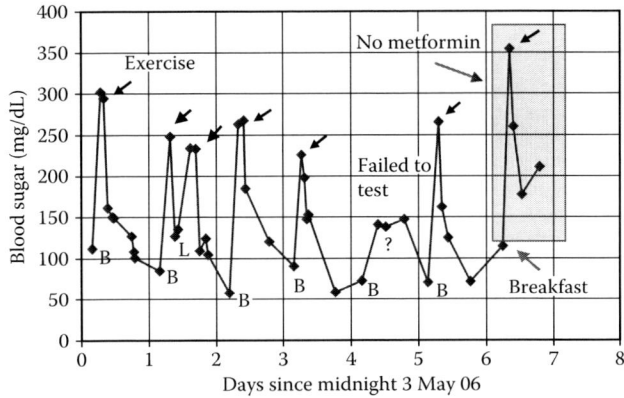

FIGURE 2.15   Effect of exercise versus metformin therapy.

of metformin: the primary means of control at that time. It is clear that the exercise was having a major effect for the entire day. Metformin was omitted on the last day, and it can be seen that the exercise had a substantially greater effect than the metformin.

2.6 Pseudocontinuum Models

Once we have been oriented at the order-of-magnitude level, it is time to develop the increasingly accurate descriptions needed to attack ever-more difficult problems. Over time, these descriptions become more complex and reliable, but they require more data and more sophisticated processing. The older and simpler models are, therefore, still useful. We suggest this by briefly reviewing the development of models for oxygenation of tissue. Development of biological systems can also follow rather simple mathematical rules that lessen the amount of information needed from the genes. We have already provided one example, the branching of arteries, and here we shall briefly discuss the fractal organization of the lungs. Finally, we briefly discuss the interaction of blood and gas flow in the lungs: a complex matching of two quite different systems that must work closely together.

2.6.1 Tissue Oxygenation

Our purpose here is to introduce and briefly describe oxygen transport and metabolism in parallel hexagonal cylinders of oxygen-demanding tissues surrounding individual capillaries. These are typical of such metabolically active systems as heart muscle or the gray matter of the brain, and modeling their behavior has received a great amount of attention [see 6–8,27]. We begin with the somewhat simplified model shown in Figure 2.16 [4, p. 35] where the hexagons have been simplified to circular cylinders as first suggested by August Krogh. One may then integrate the diffusion equations for blood and tissue. A wide variety of equations of state have been used for both blood and tissue, and it is perhaps most important to note that the metabolism of oxygen is very close to zero order. Moreover, the early simple models showed that axial dispersion and the models used for flow and diffusion within the capillaries are normally of secondary importance. This is borne out by more detailed calculations, and the profiles shown in Figure 2.16 are probably still reasonably representative of physiological operation and for general orientation [however, see 3].

They are not reliable for detailed studies however, and they fail to give a complete picture. They fail to show the interaction now known to occur between adjacent "cylinders," they require a now unnecessary geometric simplification, and they use obsolete data for parameter estimates [18,19]. The problem of selecting a modeling strategy is discussed at length by Bassingthwaighte et al. [3].

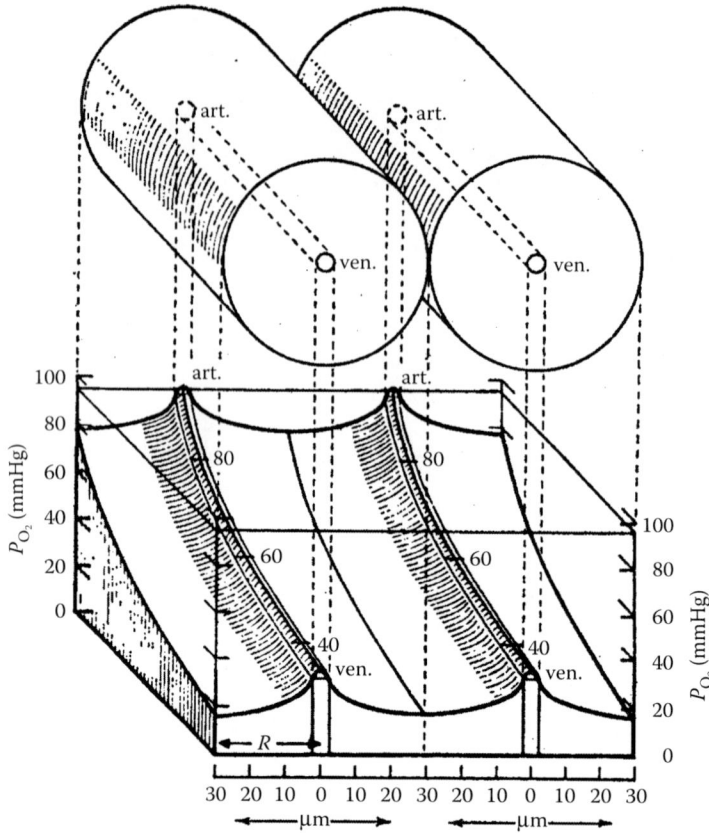

FIGURE 2.16 Tissue.

2.6.2 Pulmonary Structure and Function

We now turn to the pulmonary system sketched in Figure 2.17, a combination of convective airways and the alveolar sacs where pulmonary gases are transferred to and from the blood. As with the pulmonary blood vessels, there is a sharp transition between diffusive (in the alveoli) and convective transport in the larger airways. However, to quote Bassingthwaighte et al. [5, pp. 49–60], "the lung has two dominant features: irregularity and richness of structure." They then go on to describe this structure in some detail. It was first suggested by Mandelbrot [34] that to a good first approximation the lung is a space-filling fractal.

The diameters of the first 11 branches fall off exponentially with branch number in both the right and left lungs [5], and the flow is complex and laminar. Axial dispersion [29] is appreciable here but less than that for parabolic flow, and it decreases steadily. It is negligible for subsequent branches. There is no appreciable diffusion across these large ducts, but deposition of aerosols is quite significant, and it is currently an active area of research (Google). Size distribution is complex in the smaller airways, but shows very little species dependence [5, p. 57].

Net transport in this dead-ended system is achieved by cyclic expansion and compression of the alveolar sacs, and only a small fraction of the pulmonary gases, primarily oxygen, carbon dioxide, and water, are exchanged in a single breath. This system is much less efficient than the lungs of birds [see 58,63] or the gills of fish, but it does provide a bit of stability. The major differences in oxygen transport in each of these three vertebrates reflect in large part their evolutionary history [58]. The mammalian airway system is very close to a space-filling fractal structure [34], a fact that decreases the genetic demands in development quite significantly [57].

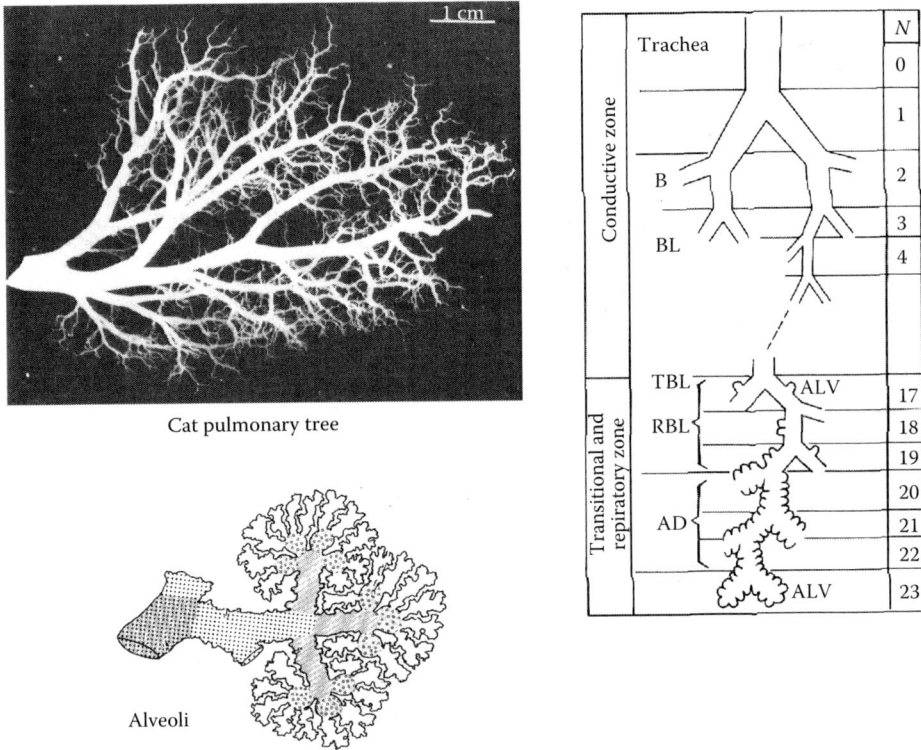

Cat pulmonary tree

Alveoli

FIGURE 2.17 Pulmonary organization oxygenation.

2.6.3 Pulmonary Blood-Gas Matching

We end our series of specific examples by noting that even in healthy individuals there is a natural tendency for blood to concentrate in the lower regions of the lungs and air in the upper. Such misdistribution can cause serious problems. This is basically a problem in process control, but its diagnosis is achieved by measuring the response of the pulmonary system to a series of volatile tracers with different solubilities in blood [e.g., 45].

2.7 More Complex Situations

It is time now to bring this discussion to a close, and it may be noted that we have proceeded from very simple processes chosen to give some appreciation for the "design strategy" on which we are based and on to less detailed descriptions of more complex processes. We are far more complex than any man-made machines, but safety factors and other "design principles" are not so different from those used by engineers. It should not, therefore, come as a surprise to find that some awkward compromises also proved necessary. We introduce just a few of them here as a reminder: many of our common medical problems arise in self-organization at three levels: individual development, species evolution, and group interactions. Some of the simpler ones include the following.

2.7.1 Stochastic Behavior of Genetic Regulation

The erratic fluctuations of gene expression can produce abnormal reaction sequences, and these can even be taken advantage of by invading pathogens. Describing these situations requires sophisticated stochastic modeling [2,28,45,51].

2.7.2 Cellular Crowding

An important related area that we unfortunately do not have space for is that of whole-organ models [e.g., 61]. It has recently been found that identical genes in apparently identical cells can do very different things. In addition to the stochastic effect just introduced is an element of noisiness that can result from crowding [22,33,42,54]. Moreover, it appears that this problem becomes worse with age.

2.7.3 The Dual Nature of Oxygen

Although the aerobic metabolisms of vertebrates confer many advantages, the formation of peroxides and other highly active oxygen compounds are extremely dangerous [31,54, Chap. 8].

2.7.4 Self-Organization and Emergence

These topics become very important in dealing with invading organisms [10,15, 20–22,26].
    Finally, we note that essentially all aspects of complexity theory find medical applications.

References

1. Alberts, B. M. et al. 2002. *The Molecular Biology of the Cell*, Garlands, New York, NY.
2. Arkin, A., J. Ross, and H. H. McAdams. 1998. Stochastic kinetic analysis of developmental pathway bifurcation in λ-infected *Escherichia coli* cells, *Genetics*, 149, 1633–1648.
3. Bassingthwaighte, J. B., H. J. Chizeck, and L. E. Atlas. 2006. Strategies and tactics in multiscale modeling of cell-to-organ systems, *Proc IEEE*, 94, 819–831.
4. Bassingthwaighte, J. B., C. Goresky, and J. Linehan. 1998. *Whole Organ Approaches to Cellular Metabolism: Permeation, Cellular Uptake and Product Formation*, Springer, New York, NY.
5. Bassingthwaighte, J. B., L. S. Liebovicth, and B. J. West. 1994. *Fractal Physiology*, Medical and Technical Publishers, Oxford.
6. Beard, D. A. 2001. The computational framework for generating transport models from data bases of microvasculature anatomy, *Ann Biomed Eng*, 29, 837–843.
7. Beard, D. A. and J. B. Bassingthwaighte. 2001. Modeling advection and diffusion of oxygen in complex vascular networks, *Ann Biomed Eng*, 29, 298–310.
8. Beard, D. A. and F. Wu. 2009. Apparent diffusivity and Taylor dispersion of water and solutes in capillary beds, *Bull Math Biol*, 71, 1366–1377.
9. Berg, O. G., and P. H. Von Hippel. 1985. Diffusion controlled macromolecular interactions, *Ann Rev Biophys Biophys Chem*, 14, 131–160.
10. Bertrand, S., J. M. Burgos, F. Gerlotti, and J. Atiquipa. 2005. Levy trajectories of Peruvian purse-seiners as an indicator of the spatial distribution of anchovy (*Engraulis ringens*), *J Marine Sci*, 62, 477–482.
11. Bird, R. B., W. E. Stewart, and E. N. Lightfoot. 2007. *Transport Phenomena*, revised 2nd ed, Wiley, New York, NY.
12. Bischoff, K. B. 1967. Applications of a mathematical model for drug distribution in mammals. In: D. Hershey (ed.), *Chemical Engineering in Medicine and Biology*, Plenum, New York, NY.
13. Bischoff, K. B., and R. L. Dedrick. 1968. Thiopental pharmacokinetics, *J. Pharma. Sci.*, 57(8):1345–1351.
14. Calder, W. A. I. 1996. *Size, Function and Life History*, Dover, Mineola, NY.
15. Camazine, S., J.-L. Deneubourg, N. F. Franks, J. Sneyd, G. Theraulaz, and E. Bonabeau. 2003, 2nd printing. *Self-Organization in Biological Systems*, Princeton Studies in Complexity, Princeton, NJ.
16. Caro, C. G., T. J. Pedley, and S. A. Seed. 1974. Mechanics of fhe circulation. In: A. C. Guyton (ed.). *Cardiovascular Physiology*, Medical and Technical Publishers, Oxford, UK.

17. Carslaw, H. S., and J. C. Jaeger. 1959. *Conduction of Heat in Solids*, 2nd ed., Oxford University Press/ Clarendon Press, Oxford, UK.

18. Dash, R. K., and J. B. Bassingthwaighte. 2006. Simultaneous blood-tissue exchange of oxygen, carbon dioxide, bicarbonate, and hydrogen iron, *Ann Biomed Eng*, 34, 1129–1148.

19. Dash, R. K., and J. B. Bassingthwaighte. 2010. Erratum to: Blood HbO_2, and $HbCO_2$ dissociation curves at varied O_2, CO_2, pH, 2.3DPG and temperature levels, *Ann Biomed Eng*, 38, 1683–1670.

20. Eigen, M. 1987. *Stufen zum Leben*, Piper, Munich.

21. Eigen, M. and R. Winkler-Oswatitsch. 1992. *Steps toward Life: A Perspective on Evolution*, Oxford University Press/Clarendon Press, Oxford, UK.

22. Ellis, R. J. 2001. Macromolecular crowding: Obvious but unappreciated. *Trends in Biomed Sci*, 26, 597–604.

23. El Samad, H., M. Khammash, L. Petsold, and D. Gillespie. 2005. Stochastic modelling of gene regulatory networks, *Int J Robust Non-Linear Contr*, 15, 691–711.

24. Feltz, B. 2006. *Self-Organization and Emergence in Life Sciences*, Springer, New York, NY.

25. Fung, Y. C. 1997. *Biomechanics: Circulation*, Springer, New York, NY.

26. Gladwell, M. 2007. Open secrets. *New Yorker* (January 8): 44.

27. Goldman, D. and A. S. Popel. 2001. A computational study of the effect of vasomotion on oxygen transport from capillary networks, *J Theor Biol*, 209, 189–199.

28. Hobbs, S. H. and E. N. Lightfoot. 1979. A Monte Carlo simulation of convective dispersion in the large airways, *Resp Physiol*, 37, 273–292.

29. Hoppensteadt, F. C. and C. H. Peskin. 2001. *Modeling and Simulation in Medicine and the Life Sciences*, 2nd ed, Springer, New York, NY.

30. Joshi, A. and B. O. Palsson. 1989. Metabolic dynamics in the human red cell. Part III. metabolic dynamics, *J Theor Biol*, 142, 41–68.

31. Lane, N. 2005. *Power, Sex and Suicide*, Oxford University Press/Clarendon Press, Oxford, UK.

32. Lane, N. 2009. *Life Ascending*, Norton/Symantec, Mountain View, CA.

33. Maamar, H., A. Raj, and D. Dubnau. 2007. Noise in gene expression determines cell fate in *Bacillus subtilis*, *Science*, 317(July), 526–529.

34. Mandelbrot, B. B. 1983. *The Fractal Geometry of Nature*, Freeman, Gordensville, VA.

35. Mastro, A. M., M. Babich, W. D. Taylor, and A. D. Keith. 1984. Diffusion of a small molecule in the cytoplasm of mammalian cells, *PNAS*, 81, 3414–3418.

36. Nesher, I, I. E. Karl, and K. M. Kipnis. 1985. Dissociation of the effect(s) of insulin and contraction on glucose transport in rat epitochlearis muscle, *Am J Physiol*, 249, C226–232.

37. Pedraza, J. M. and J. Paulsson. 2008. Effects of molecular memory and bursting on gene expression. *Science*, 319(Jan), 339–343.

38. Pereira, L. O. and A. H. Lancha. 2004. Effect of insulin and contraction upon glucose transport in skeletal muscle, *Prog Biophys Mol Biol*, 84, 1–27.

39. Pearson, H. 2008. The cellular hullabaloo, *Nature*, 453, 150–153.

40. Popel, A. S. 1989. Theory of oxygen transport to tissue, *Crit Rev Biomed Eng*, 17, 257–321.

41. Purves, M. J. 1972. *Physiology of the Cerebral Circulation*, Cambridge University Press, Cambridge.

42. Raj, A. and A. van Oudenaarden. 2009. Single molecule approaches to stochastic gene expression, *Ann Rev Biophys*, 38, 255–270.

43. Ritschel, W. and G. L. Kearns. 2004. *Handbook of Basic Pharmacokinetics*, APhA, Washington, DC.

44. Roberts, M. S., E. N. Lightfoot, and W. P. Porter. 2010 May–June. A new model for the body size-metabolism relationship, 83(3):395–405.

45. Roca, J. and P. D. Wagner. 1993. Principles and information content of the multiple inert gas elimination technique, *Thorax*, 49, 815–824.

46. Rose, A. J. and E. A. Richter. 2005. Skeletal muscle glucose uptake during exercise, *Physiology*, 20, 260–270.

47. Ruzzin, J. and J. Jensen. 2005. Contraction activates glucose uptake and glycogen synthase in muscles from dexamethazone treated rats, *Am J Physiol*, 289, E241–E250.

48. Spalding, D. B. 1958. A note on mean residence times in steady flows of arbitrary complexity, *Chem Eng Sci*, 9, 74–77.

49. Stahl, W. R. 1965. Organ weights in primates and other mammals, *Science*, 50, 1038–1042.

50. Stiles, J. 2008. *The Fundamentals of Brain Development: Integrating Nature and Nurture*, Harvard., Cambridge, MA.

51. Thattai, M. and A. van Oudenaarden. 2001. Intrinsic noise in gene regulatory networks, *PNAS*, 98, 8614–8619.

52. Turner, J. S. 2007. *The Tinkerer's Accomplice*, Harvard, Cambridge, MA.

53. Viswanathan, G. M., S. Y. Buldyrev, S. Havlin, M. G. E. da Luz, E. P. Raposo, and H. E. Stanley. 1999. Optimizing the success of random searches, *Nature,* 401, 911–914.

54. Wagner, A. 2005. *Robustness and Evolvability in Living Systems*, Princeton University Press, Princeton, NJ.

55. Wagner, P. D. 2008. The multiple inert gas elimination technique, *Intensive Care Med,* 34, 994–1001.

56. Wald, C. and C. Wu. 2010. Of mice and women: The bias in animal models, *Science,* 327, 1571–1572.

57. Warburton, D. 2008. Order in the lung. *Nature,* 453, 733–734.

58. Ward, P. D. 2006. *Out of Thin Air,* Joseph Henry Press, Washington, DC.

59. Weisz, P. B. 1973. Diffusion and reaction: An interdisciplinary excursion, *Science,* 179, 4533–440.

60. Welling, P. 1997. *Pharmacokinetics*, American Chemical Society, Washington, DC.

61. West, J. B. 2008. *Respiratory Physiology*, Lippincott, Williams and Wilkins.

62. White, C. F., and R. S. Seymour. 2003. Mammalian basic metabolism rate is proportional to body mass 2/3, *PNAS*, 100, 4046–4049.

63. Willmer, P. G. S. and I. Johnston. 2000. *Environmental Physiology of Animals*, Blackwell, Hoboken, NJ.

3

Microvascular Heat Transfer

3.1 Introduction and Conceptual Challenges 3-1
3.2 Basic Concepts 3-2
3.3 Vascular Models 3-3
    Equilibration Lengths • Countercurrent Heat Exchange • Heat
    Transfer Inside a Blood Vessel
3.4 Models of Perfused Tissues 3-6
    Continuum Models • Multiequation Models • Vascular
    Reconstruction Models
3.5 Parameter Values 3-9
    Thermal Properties • Thermoregulation • Clinical Heat
    Generation
3.6 Solutions of Models 3-11
Defining Terms 3-13
References 3-13

James W. Baish
Bucknell University

3.1 Introduction and Conceptual Challenges

Models of microvascular heat transfer are useful for optimizing thermal therapies such as hyperthermia treatment, for modeling thermoregulatory response at the tissue level, for assessing environmental hazards that involve tissue heating, for using thermal means of diagnosing vascular pathologies, and for relating blood flow to heat clearance in thermal methods of blood perfusion measurement. For example, the effect of local hyperthermia treatment is determined by the length of time that the tissue is held at an elevated temperature, nominally 43°C or higher. Since, the tissue temperature depends on the balance between the heat added by artificial means and the tissue's ability to clear that heat, an understanding of the means by which the blood transports heat is essential. This chapter of the book outlines the general problems associated with such processes while more extensive reviews and tutorials on microvascular heat transfer may be found elsewhere [1–6].

The temperature range of interest for all of these applications is intermediate between freezing and boiling, making only sensible heat exchange by conduction and convection important mechanisms of heat transfer. At high and low temperatures such as those present during laser ablation or electrocautery and cryopreservation or cryosurgery, the change of phase and accompanying mass transport present problems beyond the scope of this book. See, for example, Reference 7.

While the equations that govern heat transport are formally similar to those that govern diffusive mass transport, heat and diffusing molecules interact with the microvasculature in dramatically different ways because the thermal diffusivity of most tissues is roughly two orders of magnitude greater than the diffusivity for mass transport of most mobile species (1.5×10^{-7} m²/s for heat vs. 1.5×10^{-9} m²/s for O_2). Mass transport is largely restricted to the smallest blood vessels, the capillaries, arterioles, and venules, whereas heat transport occurs in somewhat larger, so-called thermally significant blood vessels with diameters in the range from 80 μm to 1 mm. The modes of heat transport differ from those of mass transport, not simply because these vessels are larger, but because they have a different geometrical arrangement than the vessels primarily responsible for mass transport. Many capillary beds approximate a uniformly spaced array of parallel vessels that can be well modeled by the Krogh cylinder model. In contrast, the thermally significant vessels are in a tree-like arrangement that typically undergoes several generations of branching within the size range of interest and are often found as countercurrent pairs in which the artery and vein may be separated by one vessel diameter or less. Moreover, the vascular architecture of the thermally significant vessels is less well characterized than that of either the primary mass exchange vessels or the larger, less numerous, supply vessels that carry blood over large distances in the body. There are too few supply vessels to contribute much to the overall energy balance in the tissue, but they produce large local perturbations in the tissue temperature because they are often far from thermal equilibrium with the surrounding tissue. Much of the microvascular heat exchange occurs as blood flows from the larger supply vessels into the more numerous and densely spaced, thermally significant vessels.

While the details of the vascular architecture for particular organs have been well characterized in individual cases, variability among individuals makes the use of such published data valid only in a statistical sense. Current imaging technology can be used to map and numerically model thermally significant blood vessels larger than 600 μm diameter [8], but smaller vessels must be analyzed by other approaches as illustrated later.

An additional challenge arises from the spatial and temporal variability of the blood flow in tissue. The thermoregulatory system and the metabolic needs of tissues can change the blood perfusion rates by a factor as great as 15–25.

3.2 Basic Concepts

For purposes of thermal analysis, vascular tissues are generally assumed to consist of two interacting subvolumes, a solid tissue subvolume and a blood subvolume, which contains flowing blood (see Figure 3.1). These subvolumes thermally interact through the walls of the blood vessels where heat, but

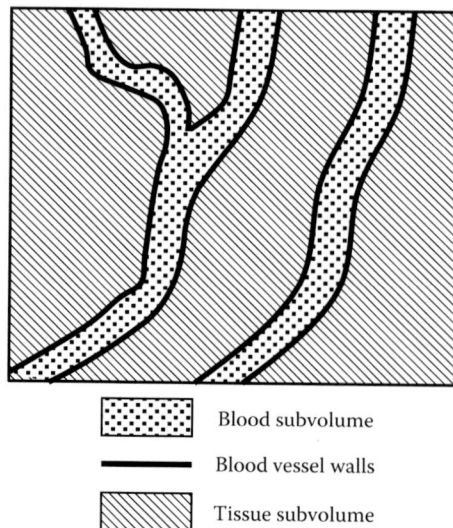

FIGURE 3.1 Schematic of blood and tissue subvolumes. The blood and tissue subvolumes are divided by the blood vessel walls with the tissue subvolume containing cells specific to the tissue, extracellular matrix, and interstitial fluid.

little mass, is exchanged. Because the tissue subvolume can transport heat by conduction alone, it may be modeled by the standard heat diffusion equation [9,10]

$$\nabla \cdot k_t \nabla T_t(\vec{r},t) + \dot{q}_t'''(\vec{r},t) = \rho_t c_t \frac{\partial T_t(\vec{r},t)}{\partial t} \tag{3.1}$$

where T_t is the local tissue temperature, k_t is the thermal conductivity of the tissue, $\dot{q}_t'''$ is the rate of volumetric heat generation from metabolism or external source, ρ_t is the tissue density, and c_t is the tissue specific heat. The properties used in Equation 3.1 may be assumed to be bulk properties that average over the details of the interstitial fluid, extracellular matrix, and cellular content of the tissue. In the blood subvolume, heat may also be transported by advection, which adds a blood velocity-dependent term as given by [9]

$$\nabla \cdot k_b \nabla T_b(\vec{r},t) - \rho_b c_b \vec{u}_b(\vec{r},t) \cdot \nabla T_b(\vec{r},t) + \dot{q}_b'''(\vec{r},t) = \rho_b c_b \frac{\partial T_b(\vec{r},t)}{\partial t} \tag{3.2}$$

where $\vec{u}_b$ is the local blood velocity and all other parameters pertain to the local properties of the blood. Potential energy, kinetic energy, and viscous dissipation effects are typically neglected.

At the internal boundary on the vessel walls, we expect a continuity of heat flux $k_b \nabla T_b(\vec{r}_w,t) = k_t \nabla T_t(\vec{r}_w,t)$ and temperature $T_b(\vec{r}_w,t) = T_t(\vec{r}_w,t)$, where $\vec{r}_w$ represents points on the vessel wall. Few attempts have been made to solve Equations 3.1 and 3.2 exactly, primarily due to the complexity of the vascular architecture and the paucity of data on the blood velocity field in any particular instance. The sections that follow present approaches to the problem of microvascular heat transport that fall broadly into the categories of vascular models that consider the response of one or a few blood vessels to their immediate surroundings and continuum models that seek to average the effects of many blood vessels to obtain a single field equation that may be solved for a local average of the tissue temperature.

3.3 Vascular Models

Most vascular models are based on the assumption that the behavior of blood flowing in a blood vessel is formally similar to that of a fluid flowing steadily in a roughly circular tube where axial conduction and heat generation in the vessel are neglected (see Figure 3.2), that is [11]

$$\pi r_a^2 \rho_b c_b \bar{u} \frac{d\bar{T}_a(s)}{ds} = q'(s) \tag{3.3}$$

where $\bar{T}_a(s)$ is the mixed mean temperature of the blood for a given vessel cross section, r_a is the vessel radius, $\bar{u}$ is the mean blood speed in the vessel, $q'(s)$ is the rate at which heat conducts into the vessel per unit length, and s is the spatial coordinate along the vessel axis. For a vessel that interacts only with a cylinder of adjacent tissue, we have

$$q'(s) = U' 2\pi r_a (\bar{T}_t(s) - \bar{T}_a(s)) \tag{3.4}$$

where U' is the overall heat transfer coefficient between the tissue and the blood. Typically, the thermal resistance inside the blood vessel is much smaller than that in the tissue cylinder, so to a first approximate, we have $U' 2\pi r_a \approx k_t \sigma$, where the conduction shape factor σ relating local tissue temperature $\bar{T}_t(s)$ to the blood temperature may be estimated from

$$\sigma \approx \frac{2\pi}{\ln(r_t/r_a)} \tag{3.5}$$

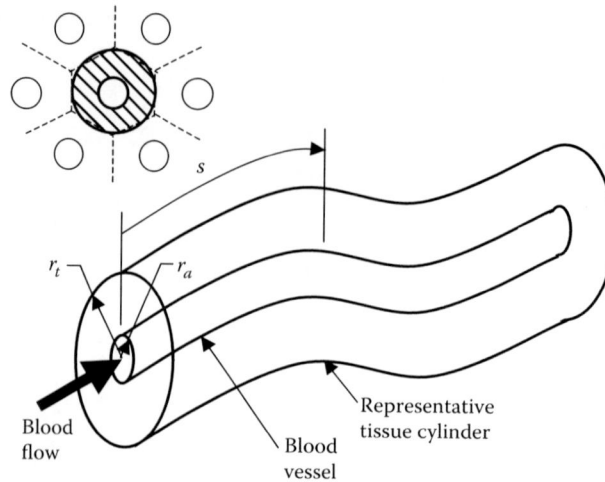

FIGURE 3.2 Representative tissue cylinder surrounding a blood vessel showing the radial and axial position coordinates. The local average tissue temperature $\bar{T}_t(s)$ is determined over the shaded, roughly circular area of solid tissue surrounding the blood vessel. The tissue cylinder has a diameter ranging from 10 to 40 times the diameter of thermally significant blood vessels.

In those instances when the thermal resistance inside the blood vessel is appreciable, it may be added in series with the resistance outside the vessel. Methods for estimating the convective heat transfer coefficient for the inside of the vessel wall are presented in Section 3.3.3.

3.3.1 Equilibration Lengths

One of the most useful concepts that arises from the simple vascular model presented earlier is the equilibration length L_e, which may be defined as the characteristic length over which the blood at an inlet temperature $\bar{T}_{a_o}$ equilibrates with surrounding tissue at a constant temperature $\bar{T}_t$. The solution for Equations 3.3 and 3.4 under these conditions is given by

$$\frac{\bar{T}_a(s) - \bar{T}_t}{\bar{T}_{a_o} - \bar{T}_t} = \exp\left(-\frac{s}{L_e}\right) \tag{3.6}$$

where the equilibration length is given by

$$L_e = \frac{\pi r_a^2 \rho_b c_b \bar{u}}{k_t \sigma} \tag{3.7}$$

Chen and Holmes [12] found that vessels with diameters of about 175 μm have an anatomical length comparable to their thermal equilibration length, thus making vessels of this approximate size the dominant site of tissue–blood heat exchange. Accordingly, these vessels are known as the *thermally significant blood vessels*. Much smaller vessels, while more numerous, carry blood that has already equilibrated with the surrounding tissue and much larger vessels, while not in equilibrium with the surrounding tissue, are too sparsely spaced to contribute significantly to the overall energy balance [11]. Even though the larger vessels do not exchange large quantities of heat with the tissue subvolume, they cannot be ignored because these vessels produce large local perturbations to the tissue temperature and form a source of blood for tissues that are at a much different temperature than the local tissue temperature.

TABLE 3.1   Shape Factors for Various Vascular Geometries

Geometry	Reference
Single vessel to skin surface	[13]
Single vessel to tissue cylinder	[12]
Countercurrent vessel to vessel	[13,16]
Countercurrent vessels to tissue cylinder	[17,18]
Countercurrent vessels with a thin tissue layer	[19]
Multiple parallel vessels	[20]
Vessels near a junction of vessels	[21]

Note: Typical dimensions of blood vessels are available elsewhere in this handbook.

3.3.2 Countercurrent Heat Exchange

Thermally significant blood vessels are frequently found in closely spaced countercurrent pairs. Only a slight modification to the preceding formulas is needed for heat exchange between adjacent arteries and veins with countercurrent flow [13]

$$q'(s) = k_t \sigma_\Delta (\bar{T}_v(s) - \bar{T}_a(s)) \tag{3.8}$$

where $\bar{T}_v(s)$ is the mixed mean temperature in the adjacent vein and the conduction shape factor is given approximately by [9]

$$\sigma_\Delta \approx \frac{2\pi}{\cosh^{-1}((w^2 - r_a^2 - r_v^2)/2r_a r_v)} \tag{3.9}$$

where w is the distance between the vessel axes and r_v is the radius of the vein. The blood temperatures in the artery and vein must be obtained simultaneously, but still yield an equilibration length of the form given in Equation 3.7. Substitution of representative property values, blood speeds, and vessel dimensions reveals that countercurrent vessels have equilibration lengths that are about one-third that of isolated vessels of similar size [13]. Based on this observation, the only vessels that participate significantly in the overall energy balance in the tissue are those larger than about 50 μm in diameter. Countercurrent exchange is sufficiently vigorous that venous blood has been observed to recapture up to 41% of that lost from artery [14].

The shape factors given earlier are only rough analytical approximations that do not include the effects of finite thermal resistance within the blood vessels and other geometrical effects. The reader is referred to Table 3.1 for references that address these issues. A careful review of the effects of the boundary condition at the vessel wall is given by Roemer [15].

3.3.3 Heat Transfer Inside a Blood Vessel

A detailed analysis of the steady-state heat transfer between the blood vessel wall and the mixed mean temperature of the blood can be done using standard heat transfer methods

$$q'(s) = h\pi d(T_w(s) - T_b) \tag{3.10}$$

where d is the vessel diameter, $T_w(s)$ is the vessel wall temperature, and the convective heat transfer coefficient h may be found from Victor and Shah's [22] recommendation that the Nusselt number may be obtained from

$$\overline{Nu}_D = \frac{hd}{k_b} = 4 + 0.155\exp(1.58\log_{10} Gz), \quad Gz < 10^3 \tag{3.11}$$

where *Gz* is the Graetz number defined as

$$Gz = \frac{\rho_b c_b \bar{u} d^2}{k_b L} \tag{3.12}$$

where *L* is the vessel length. See also Reference 23.

In the larger blood vessels, pulsatility may have pronounced effects on blood velocity and pressure. Such transient flow effects have little impact on average heat transfer rates. The Nusselt number averaged over a cycle of pulsation differs not more than 11% from the steady-state value [24]. Since the resistance to heat flow is greater in the surrounding tissue than it is within the blood vessel, the net effect of pulsatility on tissue–vessel heat transfer is generally negligible except when transients on the time scale of a cycle of pulsation are of interest.

3.4 Models of Perfused Tissues

3.4.1 Continuum Models

Continuum models of microvascular heat transfer are intended to average over the effects of many vessels so that the blood velocity field need not be modeled in detail. Such models are usually in the form of a modified heat diffusion equation in which the effects of blood perfusion are accounted for by one or more additional terms. These equations then can be solved to yield a local average temperature that does not include the details of the temperature field around every individual vessel, but provides information on the broad trends in the tissue temperature. The temperature they predict may be defined as

$$\bar{T}_t(\vec{r},t) = \frac{1}{\delta V} \int_{\delta V} T_t(\vec{r}',t) \mathrm{d}V' \tag{3.13}$$

where δV is a volume that is assumed to be large enough to encompass a reasonable number of thermally significant blood vessels, but much smaller than the scale of the tissue as a whole. Much of the confusion concerning the proper form of the bioheat equation stems from the difficulty in precisely defining such a length scale. Unlike a typical porous medium such as water percolating through sand where the grains of sand fall into a relatively narrow range of length scales, blood vessels form a branching structure with length scales spanning many orders of magnitude.

3.4.1.1 Formulations

3.4.1.1.1 *Pennes' Heat Sink Model*

In 1948, physiologist Harry Pennes modeled the temperature profile in the human forearm by introducing the assumptions that the primary site of equilibration was the capillary bed and that each volume of tissue has a supply of arterial blood that is at the core temperature of the body. The Pennes' bioheat equation has the form [25]

$$\nabla \cdot k \nabla \bar{T}_t(\vec{r},t) + \dot{\omega}_b(\vec{r},t)\rho_b c_b (T_a - \bar{T}_t(\vec{r},t)) + \dot{q}'''(\vec{r},t) = \rho c \frac{\partial \bar{T}_t(\vec{r},t)}{\partial t} \tag{3.14}$$

where $\dot{\omega}_b$ is taken to be the blood perfusion rate in volume of blood per unit volume of tissue per unit time and T_a is an arterial supply temperature that is generally assumed to remain constant and equal to the core temperature of the body, nominally 37°C. The other thermal parameters are taken to be effective values that average over the blood and tissue subvolumes. The major advantages of this formulation are that it is readily solvable for constant parameter values, requires no anatomical data, and in the absence of independent measurement of the actual blood rate and heat generation rate gives two adjustable parameters ($\dot{\omega}_b(\vec{r},t)$ and T_a) that can be used to fit the majority of the experimental results available. On the downside, the model gives no prediction of the actual details of the vascular temperatures, the actual blood perfusion rate is usually unknown and not exactly equal to the value of $\dot{\omega}_b$ that best fits the thermal data, the assumption of constant arterial temperature is not generally valid and, based on the equilibration length studies presented in the previous section, thermal equilibration occurs prior to the capillary bed. Despite these weaknesses, the Pennes' formulation is the primary choice of modelers. Equilibration prior to the capillary bed does not invalidate the model, provided that the averaging volume is large enough to encompass many vessels of the size in which equilibration actually occurs and that the venous return does not exchange significant quantities of heat after leaving the equilibrating vessels. As long as $\dot{\omega}_b$ and T_a are taken as adjustable, curve-fitting parameters rather than literally as the perfusion rate and arterial blood temperature, the model may be used fruitfully, provided that the results are interpreted accordingly.

3.4.1.1.2 Directed Perfusion

Some of the shortcomings of the Pennes' model were addressed by Wulff [26] in a formulation that is essentially the same as that used for common porous media

$$\nabla \cdot k\nabla\overline{T}_t(\vec{r},t) - \rho c\vec{u}(\vec{r},t) \cdot \nabla\overline{T}_t(\vec{r},t) + \dot{q}'''(\vec{r},t) = \rho c\frac{\partial\overline{T}_t(\vec{r},t)}{\partial t} \tag{3.15}$$

where $\vec{u}$ is a velocity averaged over both the tissue and blood subvolumes. Among the difficulties with this model are that it is valid only when the tissue and blood are in near-thermal equilibrium and when the averaging volume is small enough to prevent adjacent arteries and veins from canceling out their contributions to the average velocity, thus erroneously suggesting that the blood perfusion has no net effect on the tissue heat transfer. Equation 3.15 is rarely applied in practical situations, but served as an important conceptual challenge to the Pennes' formulation in the 1970s and the 1980s.

3.4.1.1.3 Effective Conductivity Model

The oldest continuum formulation is the effective conductivity model

$$\nabla \cdot k_{\text{eff}}\nabla\overline{T}_t(\vec{r},t) + \dot{q}'''(\vec{r},t) = \rho_t c_t\frac{\partial\overline{T}_t(\vec{r},t)}{\partial t} \tag{3.16}$$

where the effective conductivity is comprised of the intrinsic thermal conductivity of the tissue and a perfusion-dependent increment. In principle, an effective conductivity can be defined from any known heat flow and temperature difference, that is

$$k_{\text{eff}} = \frac{q}{\Delta T}f\left(\frac{L}{A}\right) \tag{3.17}$$

where $f(L/A)$ is a function of geometry with dimensions length^{-1} (e.g., $\Delta x/A$ in a slab geometry). Originally introduced as an empirical quantity [27], the effective conductivity has been linked to the Pennes' formulation in the measurement for blood perfusion rates via small, heated, implanted probes [28–30]. And in 1985, Weinbaum and Jiji [31] theoretically related the effective conductivity to the blood flow and anatomy for a restricted class of tissues and heating conditions, which are dominated by a closely spaced artery–vein architecture and which can satisfy the constraint [32]

$$\frac{d\overline{T}_t}{ds} \approx \frac{1}{2}\frac{d(\overline{T}_a + \overline{T}_v)}{ds} \tag{3.18}$$

Here, the effective conductivity is a tensor quantity related to the flow and anatomy according to [31]

$$k_{\text{eff}} = k_t \left(1 + \frac{\pi^2 \rho_b^2 c_b^2 n r_a^4 \bar{u}^2 \cos^2 \phi}{k_t^2 \sigma_\Delta} \right) \tag{3.19}$$

where the enhancement is in the direction of the vessel axes and where n is the number of artery–vein pairs per unit area and ϕ is the angle of the vessel axes relative to the temperature gradient. The near equilibrium required by Equation 3.18 is likely to be valid only in tissues in which all vessels are smaller than 200 μm diameter such as the outer few millimeters near the skin and in the absence of intense heat sources. Closely spaced, artery–vein pairs have been shown to act like highly conductive fibers even when the near equilibrium condition in Equation 3.18 is violated [33]. The radius of the thermally equivalent fiber is given by

$$r_{\text{fiber}} = (wr_a)^{1/2} \tag{3.20}$$

and its conductivity is given by

$$k_{\text{fiber}} = \frac{(\rho_b c_b \bar{u})^2 r_a^3 \cosh^{-1}(w/r_a)}{wk_t} \tag{3.21}$$

Under these nonequilibrium conditions, the tissue–blood system acts like a fiber-composite material, but cannot be well modeled as a single homogeneous material with effective properties.

3.4.1.1.4 Combination

Recognizing that several mechanisms of heat transport may be at play in tissue, Chen and Holmes [12] suggested the following formulation based on principles from porous media, which incorporates the effects discussed earlier:

$$\nabla \cdot k_{\text{eff}}(\vec{r},t)\nabla \bar{T_t}(\vec{r},t) + \dot{\omega}_b(\vec{r},t)\rho_b c_b (T_a^* - \bar{T_t}(\vec{r},t)) - \rho_b c_b \vec{u}_b(\vec{r},t) \cdot \nabla \bar{T_t}(\vec{r},t) + \dot{q}'''(\vec{r},t) = \rho_t c_t \frac{\partial T_t(\vec{r},t)}{\partial t} \tag{3.22}$$

where T_a^* is the temperature exiting the last artery that is individually modeled. The primary value of this formulation is its conceptual generality. In practice, this formulation is difficult to apply because it requires knowledge of a great many adjustable parameters, most of which have not been independently measured to date. Kahled and Vafai [34] provide an update on microvascular heat transfer from a porous media perspective.

3.4.1.1.5 Heat Sink Model with Effectiveness

Using somewhat different approaches, Brinck and Werner [35] and Weinbaum et al. [36] have proposed that the shortcomings of the Pennes' model can be overcome by introducing a heat transfer effectiveness factor ε to modify the heat sink term as follows:

$$\nabla \cdot k_t \nabla \bar{T_t}(\vec{r},t) + \varepsilon(\vec{r},t)\dot{\omega}_b(\vec{r},t)\rho_b c_b (\bar{T}(\vec{r},t)_t - T_a) + \dot{q}'''(\vec{r},t) = \rho_t c_t \frac{\partial \bar{T_t}(\vec{r},t)}{\partial t} \tag{3.23}$$

where $0 \leq \varepsilon \leq 1$. In the Brinck and Werner formulation, ε is a curve-fitting parameter that allows the actual (rather than the thermally equivalent) perfusion rate to be used [35]. Weinbaum et al. provide an analytical result for ε that is valid for blood vessels smaller than 300 μm diameter in skeletal muscle [36].

In both formulations, $\varepsilon < 1$ arises from the countercurrent heat exchange mechanism that shunts heat directly between the artery and vein without requiring the heat-carrying blood to first pass through the smaller connecting vessels. A correction factor of 0.58 is recommended for human limbs [37]. Theory predicts that the correction factor is independent of the perfusion rate.

3.4.2 Multiequation Models

The value of the continuum models is that they do not require a separate solution for the blood sub-volume. In each continuum formulation, the behavior of the blood vessels is modeled by introducing assumptions that allow solution of only a single differential equation. But by solving only one equation, all detailed information on the temperature of the blood in individual blood vessels is lost. Several investigators have introduced multiequation models that typically model the tissue, arteries, and veins as three separate, but interacting, subvolumes [13,17,38–40]. As with the other non-Pennes' formulations, these methods are difficult to apply to particular clinical applications, but provide theoretical insights into microvascular heat transfer.

3.4.3 Vascular Reconstruction Models

As an alternative to the three equation models, a more complete reconstruction of the vasculature may be used along with a scheme for solving the resulting flow, conduction, and advection equations [8,41–49]. Since, the reconstructed vasculature is similar to the actual vasculature only in a statistical sense, these models provide the mean temperature as predicted by the continuum models, as well as insight into the mechanisms of heat transport, the sites of thermal interaction, and the degree of thermal perturbations produced by vessels of a given size, but they cannot provide the actual details of the temperature field in a given living tissue. These models tend to be computationally intensive due to the high spatial resolution needed to account for all of the thermally significant blood vessels.

3.5 Parameter Values

3.5.1 Thermal Properties

The intrinsic thermal properties of tissues depend strongly on their composition. Cooper and Trezek [50] recommend the following correlations for thermal conductivity

$$k = \rho \times 10^{-3}(0.628f_{\text{water}} + 0.117f_{\text{proteins}} + 0.231f_{\text{fats}})\text{W/m-K} \tag{3.24}$$

specific heat

$$c_p = 4200f_{\text{water}} + 1090f_{\text{proteins}} + 2300f_{\text{fats}} \text{ J/kg-K} \tag{3.25}$$

and density

$$\rho = \frac{1}{f_{\text{water}}/1000 + f_{\text{proteins}}/1540 + f_{\text{fats}}/815}\text{kg/m}^3 \tag{3.26}$$

where $f_{\text{water}}, f_{\text{proteins}},$ and f_{fats} are the mass fractions of water, proteins, and fats, respectively. Representative property values are presented in Table 3.2 for tutorial purposes. A more complete tabulation is available in Reference 6. The reader is referred to the primary literature for values appropriate for specific design applications.

TABLE 3.2   Representative Thermal Property Values

Tissue	Thermal Conductivity (W/m-K)	Thermal Diffusivity (m²/s)	Perfusion (m³/m³s)
Aorta	0.461 [29]	1.25×10^{-7} [29]	
Fat of spleen	0.3337 [51]	1.314×10^{-7} [51]	
Spleen	0.5394 [51]	1.444×10^{-7} [51]	0.023 [52]
Pancreas	0.5417 [51]	1.702×10^{-7} [51]	0.0091 [52]
Cerebral cortex	0.5153 [51]	1.468×10^{-7} [51]	0.0067 [53]
Renal cortex	0.5466 [51]	1.470×10^{-7} [51]	0.077 [54]
Myocardium	0.5367 [51]	1.474×10^{-7} [51]	0.0188 [55]
Liver	0.5122 [51]	1.412×10^{-7} [51]	0.0233 [56]
Lung	0.4506 [51]	1.307×10^{-7} [51]	
Adenocarcinoma of breast	0.5641 [51]	1.436×10^{-7} [51]	
Resting muscle	0.478 [57]	1.59×10^{-7} [57]	0.0007 [55]
Whole blood (21°C)	0.492 [57]	1.19×10^{-7} [57]	
Plasma (21°C)	0.570 [57]	1.21×10^{-7} [57]	
Water	0.628 [9]	1.514×10^{-7} [9]	

Note: All conductivities and diffusivities are from humans at 37°C except the value for skeletal muscle, which is from sheep at 21°C. Perfusion values are from various mammals as noted in the references. Significant digits do not imply accuracy. The temperature coefficient for thermal conductivity ranges from −0.000254 to 0.0039 W/m-K-°C with 0.001265 W/m-K-°C typical of most tissues as compared to 0.001575 W/m-K-°C for water [51]. The temperature coefficient for thermal diffusivity ranges from $−4.9 \times 10^{-10}$ m²/s-°C to 8.4×10^{-10} m²/s-°C with 5.19×10^{-10} m²/s-°C typical of most tissues as compared to 4.73×10^{-10} m²/s-°C for water [51].

3.5.2 Thermoregulation

Humans maintain a nearly constant core temperature through a combination of physiological and behavior responses to the environment. For example, heat loss or gain at the skin surface may be modified by changes in the skin blood flow, the rate of sweating, or clothing. In deeper tissues, the dependence of the blood perfusion rate, the metabolic heat generation rate, and vessel diameters depend on the environmental and physiological conditions in a complex, organ-specific manner. The blood perfusion varies widely among tissue types, and for some tissues it can change dramatically depending on the metabolic or thermoregulatory needs of the tissue. The situation is further complicated by the feedback control aspects of the thermoregulatory systems that utilize a combination of central and peripheral temperature sensors as well as local and more distributed actuators.

The following examples are provided to illustrate some of the considerations, not to exhaustively explore this complicated issue. A model of the whole body is typically needed even for a relatively local stimulus, especially when the heat input represents a significant fraction of the whole-body heat load. The reader is referred to an extensive handbook entries on environmental response for more information [58,59]. Whole-body models of the thermoregulatory system are discussed in Reference 60.

Chato [1] suggests that the temperature dependence of the blood perfusion effect can be approximated by a scalar effective conductivity

$$k_{eff} = 4.82 - 4.44833 \left[1.00075^{-1.575(T_l - 25°C)} \right] \text{W/m-K} \tag{3.27}$$

which is intended for use in Equation 3.16.

Under conditions of local hyperthermia, where the heated volume is small compared to the body as a whole, the blood perfusion rate may undergo complex changes. Based on experimental data, the following correlations have been suggested [61,62] for muscle

$$\dot{\omega}_b \rho = \begin{cases} 0.45 + 3.55 \exp\left(-\dfrac{(T_t - 45.0°C)^2}{12.0}\right), & T_t \le 45.0°C \\ 4.00, & T_t > 45.0°C \end{cases} \tag{3.28}$$

for fat

$$\dot{\omega}_b \rho = \begin{cases} 0.36 + 0.36 \exp\left(-\dfrac{(T_t - 45.0°C)^2}{12.0}\right), & T_t \le 45.0°C \\ 0.72, & T_t > 45.0°C \end{cases} \tag{3.29}$$

and for tumor

$$\dot{\omega}_b \rho = \begin{cases} 0.833, & T_t < 37.0°C \\ 0.833 - (T_t - 37.0°C)^{4.8}/5.438 \times 10^3, & 37.0°C \le T_t \le 42.0°C \\ 0.416 & T_t > 42.0°C \end{cases} \tag{3.30}$$

Chronic heating over a period of weeks has been observed to increase vascular density and ultimately to reduce tissue temperature under constant heating conditions [63,64]. The rate and extent of adaptation are tissue specific.

The metabolic rate may also undergo thermoregulatory changes. For example, the temperature dependence of the metabolism in the leg muscle and skin may be modeled with [65]

$$\dot{q}_m''' = 170(2)^{[(T_o - T_t)/10]} \, \text{W/m}^3 \tag{3.31}$$

The metabolic rate and blood flow may also be linked through processes that reflect the fact that sustained increased metabolic activity generally requires increased blood flow.

3.5.3 Clinical Heat Generation

Thermal therapies such as hyperthermia treatment rely on local heat generation rates several orders of magnitude greater than that produced by the metabolism. Under these circumstances, the metabolic heat generation is often neglected with little error.

3.6 Solutions of Models

Numerical solutions of the Pennes' heat sink model are readily obtained by standard methods such as finite differences, finite element, boundary element, and Green's functions provided that the parameter values and appropriate boundary conditions are known [66].

Several elementary analytical solutions follow that reveal the most important scaling parameters in space and time. The steady-state solution with constant parameter values including the rate of heat generation for a tissue half space with a fixed temperature on the skin T_{skin} is given by

$$\bar{T}_t(x) = T_{\text{skin}} \exp\left[-\left(\frac{\dot{\omega}_b \rho_b c_b}{k_t}\right)^{1/2} x\right] + \left(T_a + \frac{\dot{q}'''}{\dot{\omega}_b \rho_b c_b}\right)\left\{1 - \exp\left[-\left(\frac{\dot{\omega}_b \rho_b c_b}{k_t}\right)^{1/2} x\right]\right\} \tag{3.32}$$

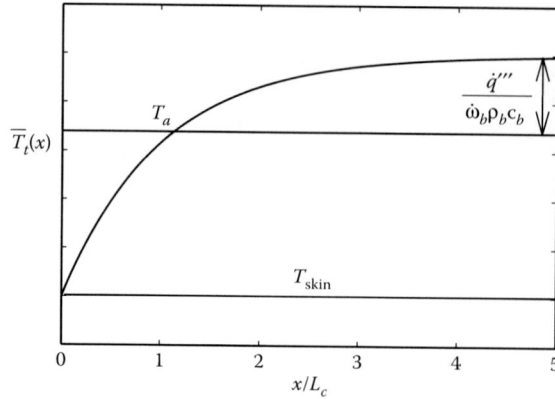

FIGURE 3.3 One-dimensional steady-state solution of Pennes' bioheat equation for constant parameter values.

This solution reveals that perturbations to the tissue temperature decay exponentially with a characteristic length of

$$L_c = \left(\frac{k_t}{\dot{\omega}_b \rho_b c_b} \right)^{1/2} \tag{3.33}$$

which for typical values of the perfusion rate $\dot{\omega}_b = 0.1 \times 10^{-3}$ to 3.0×10^{-3} m³/m³s yields $L_c = 6.5 \times 10^{-3}$ to 36×10^{-3} m (Figure 3.3).

The transient solution of Pennes' bioheat equation with constant perfusion rate for an initial uniform temperature of T_o, in the absence of any spatial dependence, is

$$\bar{T}_t(t) = T_o \exp\left[-\left(\frac{\dot{\omega}_b \rho_b c_b}{\rho_t c_t} \right) t \right] + \left(T_a + \frac{\dot{q}'''}{\dot{\omega}_b \rho_b c_b} \right) \left\{ 1 - \exp\left[-\left(\frac{\dot{\omega}_b \rho_b c_b}{\rho_t c_t} \right) t \right] \right\} \tag{3.34}$$

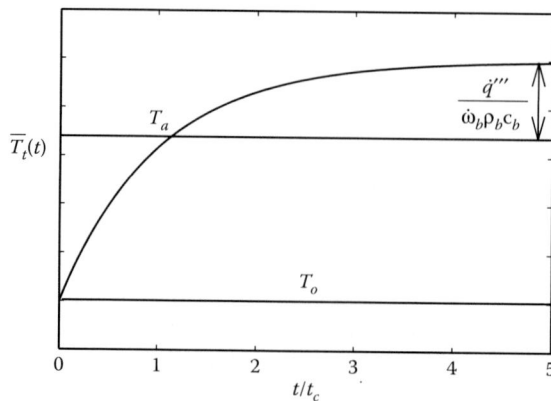

FIGURE 3.4 Transient solution of Pennes' bioheat equation for constant parameter values in the absence of spatial gradients.

Here, the solution reveals a characteristic time scale

$$t_c = \frac{\rho_t c_t}{\dot{\omega}_b \rho_b c_b} \tag{3.35}$$

that has typical values in the range of $t_c = 300$ to $10,000$ s (Figure 3.4). This solution is valid only when thermoregulatory changes in the perfusion rate are small or occur over a much longer time than the characteristic time scale t_c.

Defining Terms

Conduction shape factor: Dimensionless factor used to account for the geometrical effects in steady-state heat conduction between surfaces at different temperatures.

Effective conductivity: A modified thermal conductivity that includes the intrinsic thermal conductivity of the tissue as well as a contribution from blood perfusion effects.

Equilibration length: Characteristic length scale over which blood in a blood vessel will change temperature in response to the surrounding tissue at a different temperature.

Perfusion rate: Quantity of blood provided to a unit of tissue per unit time.

Specific heat: Quantity of energy needed to increase unit temperature for a unit of mass.

Thermal conductivity: Rate of energy transfer by thermal conduction for a unit temperature gradient per unit of cross-sectional area.

Thermally significant vessel: Blood vessels large enough and numerous enough to contribute significantly to overall heat transfer rates in tissue.

References

1. Chato, J.C., Fundamentals of bioheat transfer, in *Thermal Dosimetry and Treatment Planning*, M. Gautherie, Editor. 1990, Springer-Verlag: New York. pp. 1–56.
2. Charny, C.K., Mathematical models of bioheat transfer, in *Bioengineering Heat Transfer: Advances in Heat Transfer*, Y.I. Cho, Editor. 1992, Academic Press: Boston. pp. 19–155.
3. Arkin, M., L.X. Xu, and K.R. Holmes, Recent developments in modeling heat transfer in blood perfused tissues. *IEEE Transactions in Biomedical Engineering*, 1994. 41(2): 97–107.
4. Eto, T.K. and B. Rubinsky, Bioheat transfer, in *Introduction to Bioengineering*, S.A. Berger, W. Goldsmith, and E.R. Lewis, Editors. 1996, Oxford University Press: Oxford. pp. 203–227.
5. Minkowycz, W.J. and E.M. Sparrow, eds. *Advances in Numerical Heat Transfer*. Vol. III. 2009, CRC Press: Boca Raton, FL.
6. Diller, K.R., J.W. Valvano, and J.A. Pearce, Bioheat transfer, in *The CRC Handbook of Thermal Engineering*, F. Kreith, Editor. 2000, CRC Press: Boca Raton, FL.
7. Diller, K.R., Modeling of bioheat transfer processes at high and low temperatures, in *Bioengineering Heat Transfer: Advances in Heat Transfer*, Y.I. Cho, Editor. 1992, Academic Press: Boston. pp. 157–357.
8. Craciunescu, O.I. et al., Discretizing large traceable vessels and using DE-MRI perfusion maps yields numerical temperature contours that match the MR noninvasive measurements. *Medical Physics*, 2001. 28(11): 2289–2296.
9. Incropera, F.P. and D.P. DeWitt, *Fundamentals of Heat and Mass Transfer*. 4th ed. 1996, New York: John Wiley and Sons.
10. Incropera, F.P. et al., *Fundamentals of Heat and Mass Transfer*. 6th ed. 2007, New York: John Wiley and Sons.
11. Chato, J.C., Heat transfer to blood vessels. *Journal of Biomechanical Engineering*, 1980. 102: 110–118.

12. Chen, M.M. and K.R. Holmes, Microvascular contributions in tissue heat transfer. *Annals of the New York Academy of Sciences*, 1980. 325: 137–150.
13. Weinbaum, S., L.M. Jiji, and D.E. Lemons, Theory and experiment for the effect of vascular microstructure on surface tissue heat transfer—Part I: Anatomical foundation and model conceptualization. *Journal of Biomechanical Engineering*, 1984. 106: 321–330.
14. He, Q., D. Lemons, and S. Weinbaum, Experimental measurements of the temperature variation along artery-vein pairs from 200 to 1000 micron diameter in rat hind limb. *Journal of Biomechanical Engineering*, 2002. 124: 656–661.
15. Roemer, R.B., Conditions for equivalency of countercurrent vessel heat transfer formulations. *Journal of Biomechanical Engineering*, 1999. 121: 514–520.
16. Wissler, E.H., An analytical solution of countercurrent heat transfer between parallel vessels with a linear axial temperature gradient. *Journal of Biomechanical Engineering*, 1988. 110: 254–256.
17. Baish, J.W., P.S. Ayyaswamy, and K.R. Foster, Small-scale temperature fluctuations in perfused tissue during local hyperthermia. *Journal of Biomechanical Engineering*, 1986. 108: 246–250.
18. Shrivastava, D. and R.B. Roemer, An analytical study of 'Poisson conduction shape factors' for two thermally significant vessels in a finite, heated tissue. *Physics in Medicine and Biology*, 2005. 50(15): 3627–3641.
19. Zhu, L. and S. Weinbaum, A model for heat transfer from embedded blood vessels in two-dimensional tissue preparations. *Journal of Biomechanical Engineering*, 1995. 117(1): 64–73.
20. Cousins, A., On the Nusselt number in heat transfer between multiple parallel blood vessels. *Journal of Biomechanical Engineering*, 1997. 119(1): 127–129.
21. Baish, J.W., J.K. Miller, and M.J. Zivitz, Heat transfer in the vicinity of the junction of two blood vessels, in *Advances in Bioheat and Mass Transfer: Microscale Analysis of Thermal Injury Processes, Instrumentation, Modeling and Clinical Applications*, R.B. Roemer, Editor. 1993, ASME: New York. pp. 95–100.
22. Victor, S.A. and V.L. Shah, Steady state heat transfer to blood flowing in the entrance region of a tube. *International Journal of Heat and Mass Transfer*, 1976. 19: 777–783.
23. Barozzi, G.S. and A. Dumas, Convective heat transfer coefficients in the circulation. *Journal of Biomechanical Engineering*, 1991. 113(3): 308–313.
24. Craciunescu, O.I. and S. Clegg, Pulsatile blood flow effects on temperature distribution and heat transfer in rigid blood vessels. *Journal of Biomechanical Engineering*, 2001. 123: 500–505.
25. Pennes, H.H., Analysis of tissue and arterial blood temperatures in the resting forearm. *Journal of Applied Physiology*, 1948. 1: 93–122.
26. Wulff, W., The energy conservation equation for living tissue. *IEEE Transactions on Biomedical Engineering*, 1974. 21: 494–495.
27. Bazett, H.C. and B. McGlone, Temperature gradients in tissues in man. *American Journal of Physiology*, 1927. 82: 415–428.
28. Perl, W., Heat and matter distribution in body tissues and the determination of tissue blood flow by local clearance methods. *Journal of Theoretical Biology*, 1962. 2: 201–235.
29. Valvano, J.W. et al. *Thermal Conductivity and Diffusivity of Arterial Walls*. 1984: ASME.
30. Arkin, H., K.R. Holmes, and M.M. Chen, A technique for measuring the thermal conductivity and evaluating the "apparent conductivity" concept in biomaterials. *Journal of Biomechancial Engineering*, 1989. 111: 276–282.
31. Weinbaum, S. and L.M. Jiji, A new simplified bioheat equation for the effect of blood flow on local average tissue temperature. *Journal of Biomechanical Engineering*, 1985. 107: 131–139.
32. Weinbaum, S. and L.M. Jiji, The matching of thermal fields surrounding countercurrent microvessels and the closure approximation in the Weinbaum-Jiji equation. *Journal of Biomechanical Engineering*, 1989. 111: 271–275.
33. Baish, J.W., Heat Transport by countercurrent blood vessels in the presence of an arbitrary temperature gradient. *Journal of Biomechanical Engineering*, 1990. 112(2): 207–211.

34. Khaled, A.-R.A. and K. Vafai, The role of porous media in modeling flow and heat transfer in biological tissues. *International Journal of Heat and Mass Transfer*, 2003. 46: 4989–5003.
35. Brinck, H. and J. Werner, Efficiency function: Improvement of classical bioheat approach. *Journal of Applied Physiology*, 1994. 77: 1617–1622.
36. Weinbaum, S. et al., A new fundamental bioheat equation for muscle tissue: Part I blood perfusion term. *Journal of Biomechanical Engineering*, 1997. 119(3): 278–288.
37. Zhu, L. et al., A new fundamental bioheat equation for muscle tissue-Part II: Temperature of SAV vessels. *Journal of Biomechanical Engineering*, 2002. 124: 121–132.
38. Jiji, L.M., S. Weinbaum, and D.E. Lemons, Theory and experiment for the effect of vascular microstructure on surface tissue heat transfer-Part II: Model formulation and solution. *Journal of Biomechanical Engineering*, 1984. 106: 331–341.
39. Baish, J.W., P.S. Ayyaswamy, and K.R. Foster, Heat transport mechanisms in vascular tissues: A model comparison. *Journal of Biomechanical Engineering*, 1986. 108: 324–331.
40. Charny, C.K. and R.L. Levin, Bioheat transfer in a branching countercurrent network during hyperthermia. *Journal of Biomechanical Engineering*, 1989. 111: 263–270.
41. Baish, J.W., Formulation of a statistical model of heat transfer in perfused tissue. *Journal of Biomechanical Engineering*, 1994. 116(4): 521–527.
42. Huang, H.W., Z.P. Chen, and R.B. Roemer, A counter current vascular network model of heat transfer in tissues. *Journal of Biomechanical Engineering*, 1996. 118(1): 120–129.
43. Van der Koijk, J.F. et al., The influence of vasculature on temperature distributions in MECS interstitial hyperthermia: Importance of longitudinal control. *International Journal of Hyperthermia*, 1997. 13(4): 365–386.
44. Van Leeuwen, G.M.J. et al., Accuracy of geometrical modelling of heat transfer from tissue to blood vessels. *Physics in Medicine and Biology*, 1997. 42: 1451–1460.
45. Van Leeuwen, G.M.J., A.N.T.J. Kotte, and J.J.W. Lagendijk, A flexible algorithm for construction of 3-D vessel networks for use in thermal modeling. *IEEE Transactions on Biomedical Engineering*, 1998. 45: 596–605.
46. Kotte, A.N.T.J., G.M.J. Van Leeuwen, and J.J.K. Lagendijk, Modelling the thermal impact of a discrete vessel tree. *Physics in Medicine and Biology*, 1999. 44: 57–74.
47. Van Leeuwen, G.M.J. et al., Temperature simulations in tissue with a realistic computer generated vessel network. *Physics in Medicine and Biology*, 2000. 45: 1035–1049.
48. Raaymakers, B.W., A.N.T.J. Kotte, and J.J.K. Lagendijk, How to apply a discrete vessel model in thermal simulations when only incomplete vessel data are available. *Physics in Medicine and Biology*, 2000. 45: 3385–3401.
49. Kou, H.-S., T.-C. Shih, and W.-L. Lin, Effect of the directional blood flow on thermal dose distribution during thermal therapy: An application of a Green's function based on the porous model. *Physics in Medicine and Biology*, 2003. 48: 1577–1589.
50. Cooper, T.E. and G.J. Trezek, Correlation of thermal properties of some human tissue with water content. *Aerospace Medicine*, 1971. 42: 24–27.
51. Valvano, J.W., J.R. Cochran, and K.R. Diller, Thermal conductivity and diffusivity of biomaterials measured with self-heated thermistors. *International Journal of Thermophysics*, 1985. 6: 301–311.
52. Kapin, M.A. and J.L. Ferguson, Hemodynamic and regional circulatory alterations in dog during anaphylactic challenge. *American Journal of Physiology*, 1985. 249: H430–H437.
53. Haws, C.W. and D.D. Heistad, Effects of nimodipine on cerebral vasoconstrictor responses. *American Journal of Physiology*, 1984. 247: H170–H176.
54. Passmore, J.C., R.E. Neiberger, and S.W. Eden, Measurement of intrarenal anatomic distribution of krypton-85 in endotoxic shock in dogs. *American Journal of Physiology*, 1977. 232: H54–H58.
55. Koehler, R.C., R.J. Traystman, and M.D. Jones Jr., Regional blood flow and O_2 transport during hypoxic and CO hypoxia in neonatal and sheep. *American Journal of Physiology*, 1985. 248: H118–H124.

56. Seyde, W.C. et al., Effects of anesthetics on regional hemodynamics in normovolemic and hemor-rhaged rats. *American Journal of Physiology*, 1985. 249: H164–H173.

57. Balasubramaniam, T.A. and H.F. Bowman, Thermal conductivity and thermal diffusivity of bioma-terials: A simultaneous measurement technique. *Transactions of the ASME, Journal of Biomechanical Engineering*, 1977. 99: 148–154.

58. ASHRAE, Physiological principles and thermal comfort, in *1993 ASHRAE Handbook: Fundamentals*. 1993, ASHRAE Inc.: Atlanta, Georgia. pp. 8.1–8.29.

59. Fregly, M.J. and C.M. Blatteis, eds. *Section 4: Environmental Physiology*. Handbook of Physiology. Vol. I. 1996, American Physiological Society: New York.

60. Wissler, E.H., Mathematical simulation of human thermal behavior using whole body models, in *Heat transfer in Medicine and Biology: Analysis and Applications*, A. Shitzer and R.C. Eberhart, Editors. 1985, Plenum Press: New York. pp. 325–373.

61. Erdmann, B., J. Lang, and M. Seebass, Optimization of temperature distributions for regional hyper-thermia based on a nonlinear heat transfer model. *Annals of the New York Academy of Sciences*, 1998. 858: 36–46.

62. Lang, J., B. Erdmann, and M. Seebass, Impact of nonlinear heat transfer on temperature control in regional hyperthermia. *IEEE Transactions on Biomedical Engineering*, 1999. 46(9): 1129–1138.

63. Seese, T.M. et al., Characterization of tissue morphology, angiogenesis, and temperature in the adap-tive response of muscle tissue to chronic heating. *Laboratory Investigation*, 1998. 78(12): 1553–1562.

64. Saidel, G.M. et al., Temperature and perfusion responses of muscle and lung tissue during chronic heating in vivo. *Medical and Biological Engineering and Computing*, 2001. 39: 126–133.

65. Mitchell, J.W. et al., Thermal response of human legs during cooling. *Journal of Applied Physiology*, 1970. 29: 859–856.

66. Baish, J.W., K. Mukundakrishnan, and P.S. Ayyaswamy, Numerical models of blood flow effects in biological tissues, in *Advances in Numerical Heat Transfer*, W.J. Minkowycz and E.M. Sparrow, Editors. 2009, CRC Press: Boca Raton. pp. 29–74.

4

Fluid Dynamics for Bio Systems: Fundamentals and Model Analysis

4.1 Introduction 4-1
4.2 Elements of Theoretical Hydrodynamics 4-3
    Elements of Continuum Mechanics • Flow in Tubes
4.3 Pulsatile Flow 4-9
    Hemodynamics in Rigid Tubes: Womersley's Theory • Hemodynamics
    in Elastic Tubes • Turbulence in Pulsatile Flow
4.4 Models and Computational Techniques 4-14
    Approximations to the Navier–Stokes Equations • Computational
    Fluid Dynamics
References 4-16

Robert A. Peattie
Tufts University

Robert J. Fisher
*The SABRE Institute and
Massachusetts Institute
of Technology*

4.1 Introduction

Biological processes within living systems are significantly influenced by the flow of liquids and gases. Biomedical engineers must therefore understand hydrodynamic phenomena [1] and the vital role they play in the biological processes that occur within the body [2]. In particular, engineers are concerned with perfusion effects in the cellular microenvironment, and the ability of the circulatory and respiratory systems to provide a whole-body communication network with dynamic response capabilities. Understanding the fundamental principles of fluid flow involved in these processes is also essential for describing transport of mass and heat throughout the body, as well as understanding how tissue function can be built, reconstructed, and/or modified for clinical applications.

From a geometric and flow standpoint, the body may be considered a network of highly specialized and interconnected organ systems. The key elements of this network for transport and communication are its pathway (the circulatory system) and its medium (blood). Of interest for engineering purposes are the ability of the circulatory system to transport oxygen and carbon dioxide, glucose, other nutrients and metabolites, and signal molecules to and from the tissues, as well as to provide an avenue for stress-response agents from the immune system, including cytokines, antibodies, leukocytes, and macrophages, and system repair agents such as stem cells and platelets. The bulk transport capability provided by convective flow helps to overcome the large diffusional resistance that would otherwise be offered by such a large entity as the human body. At rest, the mean blood circulation time is of the order of 1 min. Therefore, given that the total amount of blood circulating is about 76–80 mL/kg (5.3–5.6 L for a 70 kg "standard male"), the flow from the heart to this branching network is about 95 mL/s. This and other order of magnitude estimates for the human body are available elsewhere, for example, Refs. [2–4].

Although the fluids most often considered in biofluid mechanics studies are blood and air, other fluids such as urine, perspiration, tears, ocular aqueous and vitreous fluids, and the synovial fluid in the joints can also be important in evaluating tissue system behavioral responses to induced chemical and physical stresses. For purposes of analysis, these fluids are often assumed to exhibit Newtonian behavior, although the synovial fluid and blood under certain conditions can be non-Newtonian. Since blood is a suspension, it has interesting properties; it behaves as a Newtonian fluid for large shear rates, but is highly non-Newtonian for low shear rates. The synovial fluid exhibits viscoelastic characteristics that are particularly suited to its function of joint lubrication, for which elasticity is beneficial. These viscoelastic characteristics must be accounted for when considering tissue therapy for joint injuries.

Further complicating analysis is the fact that blood, air, and other physiologic fluids travel through three-dimensional passageways that are often highly branched and distensible. Within these pathways, disturbed or turbulent flow regimes may be mixed with stable, laminar regions. For example, blood flow is laminar in many parts of a healthy circulatory system in spite of the potential for peak Reynolds numbers (defined below) of the order of 10,000. However, "bursts" of turbulence are detected in the aorta during a fraction of each cardiac cycle. An occlusion or stenosis in the circulatory system, such as the stenosis of a heart valve, will promote such turbulence. Airflow in the lung is normally stable and laminar during inspiration, but less so during expiration, and heavy breathing, coughing, or an obstruction can result in fully turbulent flow, with Reynolds numbers of 50,000 a possibility.

Although elasticity of vessel walls can significantly complicate fluid flow analysis, biologically it provides important homeostatic benefits. For example, pulsatile blood flow induces accompanying expansions and contractions in healthy elastic-wall vessels. These wall displacements then influence the flow fields. Elastic behavior maintains the norm of laminar flow that minimizes wall stress, lowers flow resistance, and thus energy dissipation, and fosters maximum life of the vessel. In combination with pulsatile flow, distensibility permits strain relaxation of the wall tissue with each cardiac cycle, which minimizes the probability of vessel failure and promotes extended "on-line" use.

The term *perfusion* is used in engineering biosciences to identify the rate of blood supplied to a unit quantity of an organ or tissue. Clearly, perfusion of *in vitro* tissue systems is necessary to maintain cell viability along with functionality to mimic *in vivo* behavior. Furthermore, it is highly likely that cell viability and normoperative metabolism are dependent on the three-dimensional structure of the microvessels distributed through any tissue bed, which establishes an appropriate microenvironment through both biochemical and biophysical mechanisms. This includes transmitting both intracellular and long-range signals along the scaffolding of the extracellular matrix.

The primary objective of this chapter is to summarize the most important concepts of fluid dynamics, as hydrodynamic and hemodynamic principles have many important applications to physiology, pathophysiology, and tissue engineering. In fact, the interaction of fluids and supported tissue is of paramount importance to tissue development and viability, both *in vivo* and *in vitro*. The strength of adhesion and dynamics of detachment of mammalian cells from engineered biomaterials and scaffolds are important subjects of ongoing research [5], as are the effects of shear on receptor–ligand binding at the cell–fluid interface. Flow-induced stress has numerous critical consequences for cells, altering transport across the cell membrane, receptor density and distribution, binding affinity and signal generation with subsequent trafficking within the cell [6]. In addition, design and use of perfusion systems such as membrane biomimetic reactors and hollow fibers is most effective when careful attention is given to issues of hydrodynamic similitude. Similarly, understanding the role of fluid mechanical phenomena in arterial disease and subsequent therapeutic applications is clearly dependent on the appreciation of hemodynamics.

Understanding of fluid phenomena is also crucial for processing and transport applications not taking place within living systems. For example, the ability to generate nanoscale entities, as in emulsions and suspensions, requires knowledge of multiphase flow and turbulent mixing concepts. Typical uses are (1) "bottom-up" drug crystal size control, (2) permeation enhancement materials for dispersion into immunoprotective barrier membranes, as in improving oxygen supply to encapsulated cells/tissue systems, and (3) creating chaperones for specific targets as in imaging and/or drug delivery. For further details and other important applications, the reader will find the following sources more appropriate [7–17].

A thorough treatment of the mathematics needed for model development and analysis is beyond the scope of this volume, and is presented in numerous sources [1,2]. Herein, the goal is to provide a physical understanding of the important issues relevant to hemodynamic flow and transport. Solution methods are summarized, and the benefits associated with use of computational fluid dynamics (CFD) packages are described. In particular, quantifying hemodynamic events can require invasive experimentation and/or extensive model and computational analysis.

4.2 Elements of Theoretical Hydrodynamics

It is essential that engineers understand both the advantages and the limitations of mathematical theories and models of biological phenomena, as well as the assumptions underlying those models. Mechanical theories often begin with Newton's second law ($\mathbf{F} = \mathbf{ma}$). When applied to continuous distributions of Newtonian fluids, Newton's second law gives rise to the *Navier–Stokes equations*. In brief, these equations provide an expression governing the motion of fluids such as air and water for which the rate of motion is linearly proportional to the applied stress producing the motion. Below, the basic concepts from which the Navier–Stokes equations have been developed are summarized along with a few general ideas about boundary layers and turbulence. Applications to the vascular system are then treated in the context of pulsatile flow. It is hoped that this very generalized approach will allow the reader to appreciate the complexities involved in an analytic solution to pulsatile phenomena, a necessity for properly describing vascular hemodynamics for clinical evaluations.

4.2.1 Elements of Continuum Mechanics

The theory of fluid flow, together with the theory of elasticity, makes up the field of continuum mechanics, the study of the mechanics of continuously distributed materials. Such materials may be either solid or fluid, or may have intermediate viscoelastic properties. Since the concept of a continuous medium, or continuum, does not take into consideration the molecular structure of matter, it is inherently an idealization. However, as long as the smallest length scale in any problem under consideration is significantly larger than the size of the molecules making up the medium and the mean free path within the medium, for mechanical purposes, all mass may safely be assumed to be continuously distributed in space. As a result, the density of materials can be considered to be a continuous function of spatial position and time.

4.2.1.1 Constitutive Equations

The response of any fluid to applied forces and temperature disturbances can be used to characterize the material. For this purpose, functional relationships between applied stresses and the resulting rate of strain field of the fluid are needed. Fluids that are homogeneous and isotropic, and for which there is a linear relationship between the state of stress within the fluid s_{ij} and the rate of strain tensor ξ_{ij}, where i and j denote the Cartesian coordinates x, y, and z, are called *Newtonian*. In physiologic settings, Newtonian fluids normally behave as if incompressible. For such fluids, it can be shown that

$$s_{ij} = -P\delta_{ij} + 2\mu\xi_{ij} \tag{4.1}$$

with μ being the dynamic viscosity of the fluid and $P = P(x, y, z)$ the fluid pressure.

4.2.1.2 Conservation (Field) Equations

In vector notation, conservation of mass for a continuous fluid is expressed through

$$\frac{\partial \rho}{\partial t} + \nabla \cdot \rho \mathbf{u} = 0 \tag{4.2}$$

where ρ is the fluid density and $\mathbf{u} = \mathbf{u}(x, y, z)$ is the vector velocity field. When the fluid is incompressible, density is constant and Equation 4.2 reduces to the well-known *continuity condition*, $\nabla \cdot \mathbf{u} = 0$. The continuity condition can also be expressed in terms of Cartesian velocity components (u, v, w) as $\partial u/\partial x + \partial v/\partial y + \partial w/\partial z = 0$.

The basic equation of Newtonian fluid motion, the Navier–Stokes equations, can be developed by substitution of the constitutive relationship for a Newtonian fluid, P-1, into the Cauchy principle of momentum balance for a continuous material [18]. In writing the second law for a continuously distributed fluid, care must be taken to correctly express the acceleration of the fluid particle to which the forces are being applied through the *material derivative Du/Dt*, where $Du/Dt = \partial \mathbf{u}/\partial t + (\mathbf{u} \cdot \nabla)\mathbf{u}$. That is, the velocity of a fluid particle may change for either of two reasons: because the particle accelerates or decelerates with time (*temporal acceleration*) or because the particle moves to a new position, at which the velocity has a different magnitude and/or direction (*convective acceleration*).

A flow field for which $\partial/\partial t = 0$ for all possible properties of the fluid and its flow is described as *steady*, to indicate that it is independent of time. However, the statement $\partial \mathbf{u}/\partial t = 0$ does not imply $D\mathbf{u}/Dt = 0$, and similarly $D\mathbf{u}/Dt = 0$ does not imply that $\partial \mathbf{u}/\partial t = 0$.

Using the material derivative, the Navier–Stokes equations for an incompressible fluid can be written in vector form as

$$\frac{D\mathbf{u}}{Dt} = \mathbf{B} - \frac{1}{\rho}\nabla P + \nu \nabla^2 \mathbf{u} \tag{4.3}$$

where ν is the fluid kinematic viscosity $= \mu/\rho$.

Expanded in full, the Navier–Stokes equations are three simultaneous, nonlinear scalar equations, one for each component of the velocity field. In Cartesian coordinates, Equation 4.3 takes the form

$$\frac{\partial u}{\partial t} + u\frac{\partial u}{\partial x} + v\frac{\partial u}{\partial y} + w\frac{\partial u}{\partial z} = B_x - \frac{1}{\rho}\frac{\partial P}{\partial x} + \nu\left(\frac{\partial^2 u}{\partial x^2} + \frac{\partial^2 u}{\partial y^2} + \frac{\partial^2 u}{\partial z^2}\right) \tag{4.4a}$$

$$\frac{\partial v}{\partial t} + u\frac{\partial v}{\partial x} + v\frac{\partial v}{\partial y} + w\frac{\partial v}{\partial z} = B_y - \frac{1}{\rho}\frac{\partial P}{\partial y} + \nu\left(\frac{\partial^2 v}{\partial x^2} + \frac{\partial^2 v}{\partial y^2} + \frac{\partial^2 v}{\partial z^2}\right) \tag{4.4b}$$

$$\frac{\partial w}{\partial t} + u\frac{\partial w}{\partial x} + v\frac{\partial w}{\partial y} + w\frac{\partial w}{\partial z} = B_z - \frac{1}{\rho}\frac{\partial P}{\partial z} + \nu\left(\frac{\partial^2 w}{\partial x^2} + \frac{\partial^2 w}{\partial y^2} + \frac{\partial^2 w}{\partial z^2}\right) \tag{4.4c}$$

Flow fields may be determined by solution of the Navier–Stokes equations, provided $\mathbf{B}$ is known. This is generally not a difficulty, since the only body force normally significant in hemodynamic applications is gravity. For an incompressible flow, there are then four unknown dependent variables, the three components of velocity and the pressure P, and four governing equations, the three components of the Navier–Stokes equations and the continuity condition. It is important to emphasize that this set of equations is *not* sufficient to calculate the flow field when the flow is compressible or involves temperature changes, since pressure, density, and temperature are then interrelated, which introduces new dependent variables to the problem.

Solution of the Navier–Stokes equations also requires that boundary conditions, and sometimes initial conditions as well, be specified for the flow field of interest. By far the most common boundary condition in physiologic and other engineering flows is the so-called *no-slip condition*, requiring that the layer of fluid elements in contact with a boundary have the same velocity as the boundary itself. For an unmoving, rigid wall, as in a pipe, this velocity is zero. However, in the vasculature, vessel walls expand and contract during the cardiac cycle.

Flow patterns and accompanying flow field characteristics depend largely on the values of governing dimensionless parameters. There are many such parameters, each relevant to specific types of flow settings, but the principle parameter of steady flows is the *Reynolds number*, Re, defined as $Re = \rho UL/\mu$, where U is a characteristic velocity of the flow field and L is a characteristic length. Both U and L must be selected for the specific problem under study, and in general, both will have different values in different problems. For pipe flow, U is most commonly selected to be the mean velocity of the flow with L being the pipe diameter.

It can be shown that the Reynolds number represents the ratio of inertial forces to viscous forces in the flow field. Flows at sufficiently low Re therefore behave as if highly viscous, with little to no fluid acceleration possible. At the opposite extreme, high Re flows behave as if lacking viscosity. One consequence of this distinction is that very high Reynolds number flow fields may at first seem to contradict the no-slip condition, in that they seem to "slip" along a solid boundary exerting no shear stress. This dilemma was first resolved in 1905 with Prandtl's introduction of the *boundary layer*, a thin region of the flow field adjacent to the boundary in which viscous effects are important and the no-slip condition is obeyed [19–21].

4.2.1.3 Turbulence and Instabilities

Flow fields are broadly classified as either *laminar* or *turbulent* to distinguish between smooth and irregular motion, respectively. Fluid elements in laminar flow fields follow well-defined paths indicating smooth flow in discrete layers or "laminae," with minimal exchange of material between layers due to the lack of macroscopic mixing. The transport of momentum between system boundaries is thus controlled by molecular action, and is dependent on the fluid viscosity.

In contrast, many flows in nature as well as engineered applications are found to fluctuate randomly and continuously, rather than streaming smoothly, and are classified as turbulent. These turbulent flows are characterized by a vigorous mixing action throughout the flow field, which is caused by *eddies* of varying size within the flow [7,14,15]. Because of these eddies fluctuate randomly, the velocity field is not constant in time. Although turbulent flows do not meet the aforementioned definition for steady, the velocity at any point presents a statistically distinct time-average value that is constant. Turbulent flows are therefore described as *stationary*, rather than truly unsteady.

Physically, the two flow states are linked, in the sense that any flow can be stable and laminar if the ratio of inertial to viscous forces is sufficiently small. Turbulence results when this ratio exceeds a *critical value*, above which the flow becomes unstable to perturbations and breaks down into fluctuations.

Fully turbulent flow fields have four defining characteristics [7,21,22]: they fluctuate *randomly*, they are *three-dimensional*, they are *dissipative*, and they are *dispersive*. The *turbulence intensity* I of any flow field is defined as the ratio of velocity fluctuations u' to time-average velocity $\bar{u}$, $I = u'/\bar{u}$.

Steady flow in straight, rigid pipes is characterized by only one dimensionless parameter, the Reynolds number. It was shown by Osborne Reynolds that for Re < 2000, incidental disturbances in the flow field are damped out and the flow remains stable and laminar. For Re > 2000, brief bursts of fluctuations appear in the velocity separated by periods of laminar flow. As Re increases, the duration and intensity of these bursts increases until they merge together into full turbulence. Laminar flow may be achieved with Re as large as 20,000 or greater in extremely smooth pipes, but it is unstable to flow disturbances and rapidly becomes turbulent if perturbed.

Since the Navier–Stokes equations govern all the behavior of any Newtonian fluid flow, it follows that turbulent flow patterns should be predictable through analysis based on those equations. However, although turbulent flows have been investigated for more than a century and the equations of motion analyzed in great detail, no general approach to the solution of problems in turbulent flow has been found. Statistical studies invariably lead to a situation in which there are more unknown variables than equations, which is called the *closure problem* of turbulence. Efforts to circumvent this difficulty have included phenomenologic concepts such as *eddy viscosity* and *mixing length* (i.e., Kolmogorov scale), as well as analytical methods that include dimensional analysis and asymptotic invariance studies [7,14,15,21,22].

4.2.2 Flow in Tubes

Flow in a tube is the most common fluid dynamic phenomenon in the physiology of living organisms, and is the basis for transport of nutrient molecules, respiratory gases, hormones, and a variety of other important solutes throughout the bodies of all complex living plants and animals. Only single-celled organisms, and multicelled organisms with small numbers of cells, can survive without a mechanism for transporting such molecules, although even these organisms exchange materials with their external environment through fluid-filled spaces. Higher organisms, needing to transport molecules and materials over larger distances, rely on organized systems of directed flows through networks of tubes to carry fluids and solutes. In human physiology, the circulatory system, which consists of the heart, the blood vessels of the vascular tree, and the "working" fluid blood, serves to transport blood throughout the body tissues. It is perhaps the most obvious example of an organ system dedicated to creating and sustaining flow in a network of tubes. However, flow in tubes is also a central characteristic of the respiratory, digestive, and urinary systems. Furthermore, the immune system utilizes systemic circulatory mechanisms to facilitate transport of antibodies, white blood cells, and lymph throughout the body, while the endocrine system is critically dependent on blood flow for delivery of its secreted hormones to the appropriate target organs or tissues. In addition, reproductive functions are also based on fluid flow in tubes. Thus, seven of the ten major organ systems depend on flow in tubes to fulfill their functions.

4.2.2.1 Steady Poiseuille Flow

The most basic state of motion for fluid in a pipe is one in which the motion occurs at a constant rate, independent of time. The pressure–flow relation for laminar, steady flow in round tubes is called *Poiseuille's law*, after J.L.M. Poiseuille, the French physiologist who first derived the relation in 1840 [23]. Accordingly, steady flow through a pipe or channel that is driven by a pressure difference between the pipe ends of just sufficient magnitude to overcome the tendency of the fluid to dissipate energy through the action of viscosity is called *Poiseuille flow.*

Strictly speaking, Poiseuille's law applies only to steady, laminar flow through pipes that are straight, rigid, and infinitely long, with uniform diameter, so that effects at the pipe ends may be neglected without loss of accuracy. However, although neither physiologic vessels nor industrial tubes fulfill all those conditions exactly, Poiseuille relationships have proven to be of such widespread usefulness that they are often applied even when the underlying assumptions are not met. As such, Poiseuille flow can be taken as the starting point for analysis of cardiovascular, respiratory, and other physiologic flows of interest.

A straight, rigid round pipe is shown in Figure 4.1, with x denoting the pipe axis and a the pipe radius. Flow in the pipe is governed by the Navier–Stokes equations, which for these conditions reduce to $d^2u/dr^2 + (1/r)(du/dr) = -\kappa/\mu$, with the conditions that the flow field must be symmetric about the pipe center line, that is, $du/dr|_{r=0} = 0$, and the no-slip boundary condition applies at the wall, $u = 0$ at $r = a$. Under these conditions, the velocity field solution is $u(r) = (\kappa/4\mu)(a^2 - r^2)$.

The velocity profile described by this solution has the familiar parabolic form known as Poiseuille flow (Figure 4.1). The velocity at the wall ($r = a$) is clearly zero, as required by the no-slip condition, while as expected on physical grounds, the maximum velocity occurs on the axis of the tube ($r = 0$) where $u_{max} = \kappa a^2/4\mu$. At any position between the wall and the tube axis, the velocity varies smoothly with r, with no step change at any point.

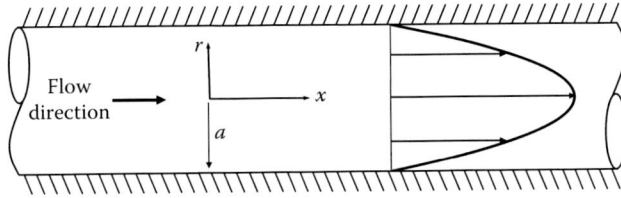

FIGURE 4.1 Parabolic velocity profile characteristic of Poiseuille flow in a round pipe of radius a. x, r—coordinate system with origin on the pipe centerline.

From physical analysis, it can be shown that the parabolic velocity profile results from a *balance* of the forces on the fluid in the pipe. The pressure gradient along the pipe accelerates fluid in the forward direction through the pipe, while at the same time, viscous shear stress retards the fluid motion. A parabolic profile is created by the balance of these effects.

Although the velocity profile is important and informative, in practice, one is therefore apt to be more concerned with measurement of the *discharge rate*, or total rate of flow in the pipe, Q, which can far more easily be accessed. The volume flow rate is given by area-integration of the velocity across the tube cross-section:

$$Q = \int_A \mathbf{u} \cdot d\mathbf{A} = \frac{\partial P}{\partial x} \frac{\pi a^4}{8\mu} \tag{4.5}$$

which is Poiseuille's law.

For convenience, the relation between pressure and flow rate is often reexpressed in an Ohm's law form, *driving force = flow × resistance*, or $\partial P/\partial x = Q \cdot (8\mu/\pi a^4)$, from which the resistance to flow, $8\mu/\pi a^4$, is seen to be inversely proportional to the fourth power of the tube radius.

A further point about Poiseuille flow concerns the area-average velocity, U. Clearly, $U = Q/\text{cross-sectional, area} = (\pi \kappa a^4/8\mu)/\pi a^2 = \kappa a^2/8\mu$. But, as was pointed out, the maximum velocity in the tube is $u_{max} = \kappa a^2/4\mu$. Hence, $U = u_{max}/2 = (1/2)u|_{r=0} = (1/2)u_{CL}$.

Finally, the shear stress exerted by the flow on the wall can be a critical parameter, particularly when it is desired to control the wall's exposure to shear. From the solution for $u(r)$, it can be shown that wall shear stress, τ_w, is given by

$$\tau_w = -\mu \frac{du}{dr} \Big|_{r=a} = \frac{\partial P}{\partial x} \frac{a}{2} = \frac{4\mu Q}{\pi a^3} \tag{4.6}$$

To summarize, Poiseuille's law, Equation 4.5, provides a relation between the pressure drop and net laminar flow in any tube, while Equation 4.6 provides a relation between the flow rate and wall shear stress. Thus, physical forces on the wall may be calculated from knowledge of the flow fields.

4.2.2.2 Entrance Flow

It can be shown that a Poiseuille velocity profile is the velocity distribution that minimizes energy dissipation in steady laminar flow through a rigid tube. Consequently, it is not surprising that if the flow in a tube encounters a perturbation that alters its profile, such as a branch vessel or a region of stenosis, immediately downstream of the perturbation the velocity profile will be disturbed away from a parabolic form, perhaps highly so. However, if the Reynolds number is low enough for the flow to remain stable as it convects downstream from the site of the original distribution, a parabolic form is gradually recovered. Consequently, at a sufficient distance downstream, a fully developed parabolic velocity profile again emerges.

Both blood vessels and bronchial tubes of the lung possess an enormous number of branches, each of which produces its own flow disturbance. As a result, many physiologic flows may not be fully developed over a significant fraction of their length. It therefore becomes important to ask, what length of tube is required for a perturbed velocity profile to recover its parabolic form, that is, how long is the entrance length in a given tube? This question can be formally posed as: if x is the coordinate along the tube axis, for what value of x does $u|_{r=0} = 2U$? Through dimensional analysis it can be shown that $x/d = const \times (\rho\ dU/\mu) = const \times Re$, where d is the tube diameter. Thus, the length of tube over which the flow develops is $const \times Re \times d$. The constant must be determined by experiment, and is found to be in the range 0.03–0.04.

Since the entrance length, in units of tube diameters, is proportional to the Reynolds number and the mean Reynolds number for flow in large tubes such as the aorta and trachea is of the order of 500–1000, the entrance length in these vessels can be as much as 20–30 diameters. In fact, there are few segments of these vessels even close to that length without a branch or curve that perturbs their flow. Consequently, flow in these vessels can be expected to almost never be fully developed. In contrast, flow in the smallest bronchioles, arterioles, and capillaries may take place with $Re < 1$. As a result, their entrance length is $\ll 1$ diameter, and flow in them will virtually always be nearly or fully developed.

4.2.2.3 Mechanical Energy Equation

Flow fields in tubes with more complex shapes than simple straight pipes, such as those possessing bends, curves, orifices, and other intricacies, are often analyzed with an *energy balance* approach, since they are not well described by Poiseuille's law. Understanding such flow fields is important to establish dynamic similitude parameters for *in vitro* studies and perfusion devices, as well as for *in vivo* studies of curved and/or branched vessel flows. For any system of total energy E, the *first law of thermodynamics* states that any change in the energy of the system ΔE must appear as either heat transferred to the system in unit time Q or as work done by the system W, so that $\Delta E = Q - W$. Here a sign convention is taken such that Q, when positive, represents heat transferred *to* the system and W, when positive, is the work done *by* the system on its surroundings. The general form of the energy equation for a fluid system is

$$\dot{Q} - \dot{W}_s = \frac{d}{dt}\int_V \left(\frac{U^2}{2} + gz + e\right)\rho dV + \int_S \left(\frac{p}{\rho} + \frac{U^2}{2} + gz + e\right)\rho u \cdot dS \tag{4.7}$$

where W_s, the "shaft work," represents work done on the fluid contained within a volume V bounded by a surface S by pumps, turbines, or other external devices through which power is often transmitted by means of a shaft; $U^2/2$ is the kinetic energy per unit mass of the fluid within V, gz is its potential energy per unit mass, with z the vertical coordinate and g gravitational acceleration, e is its internal energy per unit mass, and the density ρ is assumed to be constant.

The general equation can be simplified greatly when the flow is steady, since the total energy contained within any prescribed volume is then constant, and $d/dt = 0$. Applying Equation 4.7 to steady flow through a control volume whose end faces are denoted 1 and 2, respectively, then gives

$$\frac{p_1}{\gamma} + \beta_1\frac{U_1^2}{2g} + z_1 + h_p = \frac{p_2}{\gamma} + \beta_2\frac{U_2^2}{2g} + z_2 + h_L \tag{4.8}$$

where p_1 and p_2 are the pressures at faces 1 and 2, z_1 and z_2 are the vertical positions of those faces, $\gamma = \rho g$, h_p represents head supplied by a pump, and the coefficients β_1 and β_2 are kinetic energy correction factors introduced to simplify notation. Calculations show that $\beta = 1$ when the velocity is uniform across the section and $\beta = 2$ for laminar Poiseuille flow. Mechanical energy lost from the system is lumped together as a single term called *head loss*, h_L. For flow in a rigid pipe of length L and diameter d, h_L is well represented by $h_L = f(L/d)(U^2/2g)$, where f is called the *friction factor* of the pipe, and depends on both the pipe roughness and the flow Reynolds number. It can be shown that for laminar flow,

$f = 64/\text{Re}$. Then, $h_L = (32\mu \ LU/\gamma d^2)$. Forms that h_L can take on in turbulent flows are given in a variety of texts [24,25].

It is worth repeating that Equation 4.8 is only correct when the fluid density is constant, as is normally the case in tissue and engineering applications and even for air flow in the lung. Compressibility effects require separate energy considerations.

4.3 Pulsatile Flow

Flow in a straight, round tube driven by an axial pressure gradient that varies in time is the basis for blood transport in the arterial tree as well as respiratory gas transport in the trachea and bronchi. When the flow is confined within a tube of *rigid*, undeformable walls, its direction will always be parallel to the tube axis, so that there will only be an axial component of velocity $\mathbf{u} = (u(r,t),0,0)$ (Figure 4.2). Since all the fluid elements in the tube will then respond to any change in the pressure magnitude instantaneously and in unison, regardless of axial position, the velocity profile will be the same at all positions along the tube. It is as if all the fluid in the tube moves as a single rigid body.

As a result of the flow field acceleration and decelerations in pulsatile flows, a special type of boundary layer known as the *Stokes layer* develops. When the pressure gradient varies sinusoidally in time, as the pressure increases to its maximum, the flow increases, and as the pressure decreases, the flow does also. If the oscillations are of very low frequency, the velocity field will essentially be in phase with the pressure gradient and the boundary layers will have adequate time to grow into the tube core region. In the limit of very low frequency, the velocity field must therefore approach that of a steady Poiseuille flow. As frequency increases, however, the pressure gradient changes more rapidly and the flow begins to lag behind it due to the inertia of the fluid. The Stokes layers then become confined to a region near the wall, lacking the time required for further growth. In addition, the flow amplitude decreases with increasing oscillation frequency, as pressure gradient reversals occur more and more rapidly. In the limit of very high frequency, fluid in the tube center hardly moves at all and the Stokes layers are confined to a very thin region along the wall.

Because of the inertia of the fluid, the Stokes layer thickness, δ, is inversely related to the flow frequency, with $\delta \propto (\nu/\omega)^{1/2}$, where ω is the flow angular frequency (in rad/s).

4.3.1 Hemodynamics in Rigid Tubes: Womersley's Theory

The rhythmic contractions of the heart produce a pressure distribution in the arterial tree that includes both a steady component, P_s, and a purely oscillatory component, P_{osc}, as does the velocity field.

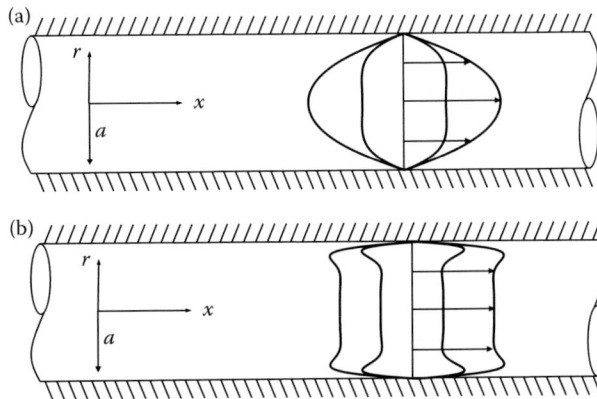

FIGURE 4.2 Representative velocity profiles of laminar, oscillatory flow in a straight, rigid tube, at four phases of the flow cycle. (a) $a = 3$, (b) $a = 13$.

In contrast, flow in the trachea and bronchi has no steady component, and thus is purely oscillatory. It is common practice to refer to these components of pressure and flow as *steady* and *oscillatory*, respectively, and to use the term *pulsatile* to refer to the superposition of the two. A very useful feature of these flows, when they occur in rigid tubes, is that the governing equation (Equation 4.10) is linear, since the flow field is unidirectional and independent of axial position. The steady and oscillatory components can therefore be decoupled from each other, and analyzed separately. This gives

$$P(x,t) = P_s(x) + P_{osc}(x,t)$$
$$u(r,t) = u_s(r) + u_{osc}(r,t)$$

(4.9)

The oscillatory component of this flow may be analyzed assuming the flow to be fully developed, so that entrance effects may be neglected, and to be driven by a purely oscillatory pressure gradient, $-(1/\rho)(\partial P/\partial x) = K\cos(\omega t) = \text{Re}(Ke^{i\omega t})$, where $i = \sqrt{-1}$ and here "Re" indicates the Real part of $Ke^{i\omega t}$. It is also convenient to introduce a new dimensionless parameter, the Womersley number, α [26], defined as $\alpha = a(\omega/\nu)^{1/2}$. Thus defined, α represents the ratio of the tube radius to the Stokes layer thickness.

The velocity field is then governed by

$$\frac{\partial u}{\partial t} = -\frac{1}{\rho}\frac{\partial P}{\partial x} + \upsilon\left(\frac{\partial^2 u}{\partial r^2} + \frac{1}{r}\frac{\partial u}{\partial r}\right)$$

(4.10)

subject to the no-slip boundary condition at the tube wall, which for a round tube takes the form $\mathbf{u} = 0$ for $r = a$.

The particular solution to equation 4.10 under this condition is most easily expressed in terms of complex *ber* and *bei* functions, which themselves are defined through [27] $ber(r) + i \cdot bei(r) = J_0(r \cdot i\sqrt{i})$, where J_0 represents the complex Bessel function of the first kind. Then

$$u(r,t) = \frac{K}{\omega}(B\cos\omega t + (1 - A)\sin\omega t)$$

(4.11)

where

$$A = \frac{ber\alpha \cdot ber\alpha\dfrac{r}{a} + bei\alpha \cdot bei\alpha\dfrac{r}{a}}{ber^2\alpha + bei^2\alpha}$$

(4.12a)

and

$$B = \frac{bei\alpha \cdot ber\alpha\dfrac{r}{a} - ber\alpha \cdot bei\alpha\dfrac{r}{a}}{ber^2\alpha + bei^2\alpha}$$

(4.12b)

Representative velocity profiles derived from these expressions are shown in Figure 4.2 for two values of α, at four phases of the flow cycle. In these figures, the radial position, r, has been normalized by the tube radius, a. At $\alpha = 3$ (Figure 4.2a), a value that under resting conditions can occur in the smallest arteries and larger arterioles as well as the middle airways, Stokes layers can occupy a significant fraction of the tube radius. The velocity at the wall is zero, as required by the no-slip condition, and as in steady flow the velocity varies smoothly with r, with no step change at any point. However, even at this low α, the velocity profile resembles a parabola only during peak flow rates. At other flow phases, a more uniform profile forms across the tube core.

In contrast, at $\alpha = 13$ (Figure 4.2b), which characterizes rest state flow in the aorta and trachea, the velocity profile of the pipe core is nearly uniform at all flow phases. Flow in the boundary layer is out of phase with that in the core, and flow reversals are possible in the Stokes layer. These changes in the velocity fields result from the inertia of the fluid, since as the flow frequency increases, less time is available in each flow cycle to accelerate the fluid.

To these flow fields of course must be added a steady component if the flow field is pulsatile rather than purely oscillatory.

As with steady flows, it is important to be able to use these expressions for the velocity field to determine the instantaneous total volume flow rate, Q_{inst}, or equivalently the instantaneous mean velocity, U_{inst}, since $Q_{inst} = U_{inst} \times$ pipe area. Following Reference 28, it can be shown that the mean velocity is

$$U(t) = \frac{K}{\omega}\left(\frac{2D}{\alpha}\cos\omega t + \left(\frac{1 - 2C}{\alpha}\right)\sin\omega t\right)$$

$$= \frac{K}{\omega}\sigma\cos(\omega t - \delta) \tag{4.13}$$

where

$$C = \frac{ber\alpha \cdot bei'\alpha - bei\alpha \cdot ber'\alpha}{ber^2\alpha + bei^2\alpha} \tag{4.14a}$$

$$D = \frac{ber\alpha \cdot ber'\alpha + bei\alpha \cdot bei'\alpha}{ber^2\alpha + bei^2\alpha} \tag{4.14b}$$

$$\sigma^2 = \left(\frac{1 - 2C}{\alpha}\right)^2 + \left(\frac{2D}{\alpha}\right)^2 \tag{4.14c}$$

$$\tan\delta = \frac{(1 - 2C/\alpha)}{(2D/\alpha)}. \tag{4.14d}$$

The oscillatory shear stress at the wall, $\tau_{w,osc}$, is given by $\tau_{w,osc} = -\mu \left(\partial u_{osc}/\partial r\right)|_{r=a}$. This results in

$$\tau_{w,osc} = \text{Re}\left(\frac{\rho K a \sqrt{i}}{\alpha} \frac{J_1\left(a\sqrt{-(i\omega/\upsilon)}\right)}{J_0\left(a\sqrt{-(i\omega/\upsilon)}\right)} \cdot e^{i\omega t}\right). \tag{4.15}$$

As with the oscillatory flow rate, the oscillatory wall shear stress lags the pressure gradient, reaching a maximum during peak flow.

4.3.2 Hemodynamics in Elastic Tubes

Because of the mathematical complexity of analysis of pulsatile flows in elastic tubes, and the variety of physical phenomena associated with them, space does not permit a full description of this topic. The reader is instead referred to a number of excellent sources for a more complete treatment [29–31]. Here we only briefly summarize the most important features of these flows, to give the reader a sense of the richness of the physics underlying them.

In brief, in a tube with a nonrigid wall, any pressure change within the tube will lead to localized bulging of the tube wall in the high pressure region (Figure 4.3). Fluid can then flow in the radial direction into the bulge. Hence, not only is the radial velocity v no longer zero, but both u and v can no longer be independent of x even far from the tube ends. Thus the flow field is governed by the continuity condition along with the full Navier–Stokes equations. Assuming axial symmetry of the tube, these become

$$\frac{\partial u}{\partial x} + \frac{\partial v}{\partial r} + \frac{v}{r} = 0 \tag{4.16}$$

$$\frac{\partial u}{\partial t} + u\frac{\partial u}{\partial x} + v\frac{\partial u}{\partial r} = -\frac{1}{\rho}\frac{\partial P}{\partial x} + \nu\left(\frac{\partial^2 u}{\partial x^2} + \frac{\partial^2 u}{\partial r^2} + \frac{1}{r}\frac{\partial u}{\partial r}\right)$$

$$\frac{\partial v}{\partial t} + u\frac{\partial v}{\partial x} + v\frac{\partial v}{\partial r} = -\frac{1}{\rho}\frac{\partial P}{\partial r} + \nu\left(\frac{\partial^2 v}{\partial x^2} + \frac{\partial^2 v}{\partial r^2} + \frac{1}{r}\frac{\partial v}{\partial r} - \frac{v}{r^2}\right) \tag{4.17}$$

The most important consequence of this is that even if the inlet pressure gradient depends only on t, within the tube the pressure gradient depends on x as well as t. An oscillatory pressure gradient applied at the tube entrance therefore propagates down the tube in a wave motion. Both the pressure and the velocity fields therefore take on wave characteristics.

The speed with which these waves travel down the tube can be expected to depend on the fluid inertia, that is, on its density, and on the wall stiffness. If the wall thickness is small compared to the tube radius and the effect of viscosity is neglected, the wave speed c_0 is given by the Moen–Korteweg formula $c_0 = (Eh/\rho d)^{1/2}$, where E is the stiffness, or Young's modulus, of the tube wall and h is its thickness. As can be expected on physical grounds, the wave speed increases as the wall stiffness rises until when E becomes infinite, the wall is rigid. Thus, oscillatory motion in a *rigid* tube, in which all the fluid moves together in bulk, may be thought of as resulting from a wave traveling with infinite speed, so that any change in the pressure gradient is felt throughout the whole tube instantaneously. In an *elastic* tube, by contrast, pressure changes are felt locally at first and then propagate downstream at finite speed.

Because of the action of the pressure and shear stress on the wall position and displacement, oscillatory flow in an elastic tube is inherently a *coupled* problem, in the sense that it is not possible in general to determine the fluid motion without also determining the resulting wall motion; the two are intrinsically linked. It can be shown [29] that the motion of the *wall* is governed by

$$\frac{\partial^2 \varsigma}{\partial t^2} = \frac{E}{(1-\sigma^2)\rho_w}\left(\frac{\partial^2 \varsigma}{\partial x^2} + \frac{\sigma}{a}\frac{\partial^2 \eta}{\partial x}\right) - \frac{\tau_w}{\rho_w h} \tag{4.18a}$$

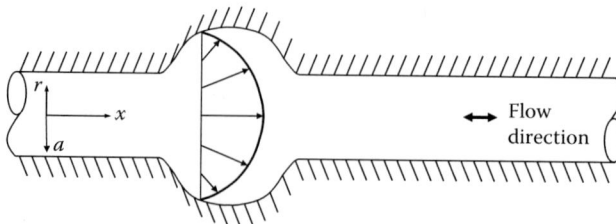

FIGURE 4.3   Local bulging of the tube wall at regions of high pressure in pulsatile flow in an elastic tube.

$$\frac{\partial^2 \eta}{\partial t^2} = \frac{P_w}{\rho_w h} - \frac{E}{(1 - \sigma^2)\rho_w a}\left(\frac{\eta}{a} + \sigma\frac{\partial \zeta}{\partial x}\right) \tag{4.18b}$$

where ζ and η are the axial and radial displacement of the wall, respectively (both of which may vary with axial position x), P_w and τ_w are the fluid pressure and shear stress at the wall, ρ_w is the wall density, and σ is the Poisson's ratio, a wall material property. Equations 4.18a and b indicate the coupling of the wall and fluid motions, since they explicitly describe ζ and η, which are properties of the *wall*, in terms of P_w and τ_w, which are themselves properties of the *flow*. In addition, coupling is imposed by the no-slip boundary condition, since the layer of fluid in contact with the wall must have the same velocity as the wall. Hence, $(\partial \zeta/\partial t) = u(x,a,t)$, the axial component of velocity at the wall, and $(\partial \eta/\partial t) = v(x,a,t)$, the radial component of velocity at the wall.

With these governing equations and boundary conditions in place, and if the input pressure distribution that drives the flow field is known, it is possible to develop a formal solution for the axial velocity. For an oscillatory flow, the input pressure would normally be expected to be of a sinusoidal form $P(x,r,t) = \text{const} \cdot e^{i\omega t}$. Following Reference 29, the method of characteristics shows the pressure distribution throughout the tube to be $P(x,r,t) = A(x,r)e^{i\omega(t-x/c)}$, where c is the wave speed in the fluid and A is the pressure amplitude. Since the fluid must be taken to be viscous, c is not equal to c_0, the inviscid fluid wave speed. Instead, $c = c_0\,(2/(1-\sigma^2)z)^{1/2}$, with z a parameter of the problem that depends on a, ω, v, σ, ρ, ρ_w, and h. It can also be shown that the pressure amplitude A depends on x, but not on r, and therefore the pressure is uniform across any axial position in the tube [29]. Under these conditions, the solution for u, the principal velocity component of interest, can be stated as

$$u(x,r,t) = \text{Re}\left(\frac{A}{\rho c}\left\{1 - G\frac{J_0\left(r\sqrt{-(i\omega/\upsilon)}\right)}{J_0\left(a\sqrt{-(i\omega/\upsilon)}\right)}\right\} \cdot e^{i\omega(t-x/c)}\right) \tag{4.19}$$

with G a factor that modifies the velocity profile shape compared to that in a rigid tube due to the wall elasticity. G is given by

$$G = \frac{2 + z(2v - 1)}{z(2v - g)} \tag{4.20}$$

with

$$g = \frac{2J_1\left(a\sqrt{-(i\omega/\upsilon)}\right)}{\left(a\sqrt{-\frac{i\omega}{\upsilon}}\right)J_0\left(a\sqrt{-(i\omega/\upsilon)}\right)}. \tag{4.21}$$

It is apparent from inspection of Equation 4.19 that the difference between the velocity field in a rigid tube and that in an elastic tube is contained in the factor G. However, since G is complex, and both its real and imaginary parts depend on the flow frequency ω, the difference is by no means readily evident. The reader is referred to Reference 29 for detailed depiction of representative velocity profiles. Nevertheless, it is important to note here that because the pressure distribution in an elastic tube takes the form of a traveling wave, *two separate* periodic oscillations can be derived from Equation 4.19. The first is that at any given axial position in the tube, the velocity profile varies sinusoidally with time, just as it does in a rigid tube. The second, however, is that at any instant of time during the flow cycle, the velocity field also varies sinusoidally in space. Fluid flows *away* from regions in which the pressure is greatest and *toward* regions in which it is least. In a rigid tube, there is only one region of maximum pressure, the upstream tube end, and only one region of minimum pressure, the downstream end. Between them, the pressure varies linearly with axial position x. In contrast, in an elastic tube, the pressure varies *sinusoidally* with x, so that many high pressure regions can exist along the tube, and these lead to a series of flow reversals at any specific time.

A final word about oscillatory flow in an elastic tube concerns the possibility of *wave reflections*. In a rigid tube, there is no wave motion as such, and flow arriving at an obstruction or branch is disturbed in some way, but otherwise progresses through the obstruction. In contrast, the wave nature of flow in an elastic tube leads to entirely different behavior at an obstacle. At an obstruction such as a bifurcation or a branch, some of the energy associated with pressure and flow is transmitted through the obstruction, while the remainder is *reflected*. This leads to a highly complex pattern of superposing primary and reflected pressure and flow waves, particularly in the arterial tree since blood vessels are elastic and vessel branchings are ubiquitous throughout the vascular system. Such wave reflections may be analyzed in terms of transmission line theory [18,29].

4.3.3 Turbulence in Pulsatile Flow

Transition to turbulence in oscillatory pipe flows occurs through fundamentally different mechanisms than transition in steady flows, for two reasons. The first is that the oscillatory nature of the flow leads to a unique base state, the most important feature of which is the formation of an oscillatory Stokes layer on the tube wall. This layer has its own stability characteristics, which are not comparable to the stability characteristics of the boundary layer of steady flow. The second reason is that temporal deceleration destabilizes the whole flow field, so that perturbations of the Stokes layer can cause the flow to break down into unstable, random fluctuations. Instability often occurs during the deceleration phase of the flow cycle, and is immediately followed by relaminarization as the net flow decays to zero prior to reversal. Because of these characteristics, during deceleration phases of the flow cycle instabilities are observable in the Stokes layer even at much lower Reynolds numbers than those for which they would be found in steady flow [32].

Since the Stokes layer thickness δ itself depends on the flow frequency, transition to turbulence depends on the Womersley number as well as the Reynolds number. Experimental measurements of the velocity made in rigid tubes by noninvasive optical techniques [32] have shown that over a range of values of $\alpha \geq 8$, the flow was found to be fully laminar for $Re_\delta \leq 500$, where Re_δ is the Reynolds number based on the Stokes layer thickness rather than tube diameter. That is, $Re_\delta = U\delta/\nu$. For $500 < Re_\delta < 1300$, the core flow remained laminar while the Stokes layer became unstable during the deceleration phase of fluid motion. This turbulence was most intense in an annular region near the tube wall. These results are in accord with theoretical predictions of instabilities in Stokes layers [33,34]. For higher values of Re_δ, instability can be expected to spread across the tube core.

4.4 Models and Computational Techniques

4.4.1 Approximations to the Navier–Stokes Equations

The Navier–Stokes equations, Equations 4.3 and 4.4, together with the continuity condition, provide a complete set of governing equations for the motion of an incompressible Newtonian fluid. If appropriate boundary and initial conditions can be specified for the motion of such a fluid in a given flow system, in principle, a full set of governing equations and conditions for the system will be known. It may then be expected that the fluid motion can be deduced simply by solution of the resulting boundary value problem. Unfortunately, however, the mathematical difficulties resulting from the nonlinear character of the acceleration terms $D\mathbf{u}/Dt$ in the Navier–Stokes equations are so great that only a very limited number of exact solutions have ever been found. The simplest of these pertain to cases in which the velocity has the same direction at every point in the flow field, as in the steady and pulsatile pipe flows discussed earlier.

Accordingly, there is a strong incentive to seek conditions under which one or more of the terms in Equation 4.3 are negligible or nearly so, and therefore an approximate and much simpler governing equation can be generated by neglecting them altogether. For example, the Reynolds number represents the ratio of inertial to viscous forces in the flow field. Accordingly, in flows for which $Re \gg 1$, it can be shown that the viscous term $v \nabla^2 \mathbf{u}$ is very much smaller than the acceleration $D\mathbf{u}/Dt$. Consequently, it can be omitted from the governing equation, which leads to solutions that are approximately valid at least outside the boundary layer. Conversely, when $Re \ll 1$, the viscous term $v \nabla^2 \mathbf{u}$ is much larger than the acceleration $D\mathbf{u}/Dt$.

In summary, these approximations show that viscosity is important in three situations:

1. When the overall Reynolds number is *low*, since then viscous effects act over the full flow field.
2. When the overall Reynolds number is *high*, since then viscosity is important in thin boundary layers.
3. When the flow is *enclosed*, as in a pipe flow, since then the available diffusion time is very large, and viscous effects can become important in the whole flow after some initial region or time.

An alternative approach to seeking simplifications to the Navier–Stokes equations is to accept the full set of equations, but approximate each term in the equation with a simpler form that permits solutions to be developed. Although the resulting equations are only *approximately* correct, the advent of modern digital computers has allowed them to be written with great fineness, so that highly accurate solutions are achieved. These techniques are called *computational fluid dynamics* (CFD).

4.4.2 Computational Fluid Dynamics

The steady improvement in computer speed and memory capabilities since the 1950s has made it possible for CFD to become a very powerful and versatile tool for the analysis of complex problems of interest in the engineering biosciences. By providing a cost-effective means to simulate real flows in detail, CFD permits studying complex problems combining thermodynamics, chemical reaction kinetics, and transport phenomena with fluid flow aspects. In addition, such problems often arise in highly complex geometries. Consequently, they may be far too difficult to study accurately without computational model approaches [35,36].

Furthermore, CFD offers a means for testing flow conditions that are unachievable or prohibitively expensive to test experimentally. For example, most flow loops and wind tunnels are limited to a fixed range of flow rates and governing parameter values. Such limits generally do not apply to CFD analyses. Moreover, flow under a wide range of parameter values may be tested with far less cost than performing repeated experiments.

A representative example of widespread interest to biomedical engineers is the analysis of hemodynamics in blood vessel models. When analyzing biologic responses to flow, or before employing newly developed surgical procedures, characterization studies need to be conducted to substantiate applicability. Cellular metabolic rates in encapsulated and free states, as well as pertinent transport phenomena, can be evaluated in anatomically realistic vessel configurations. These data, coupled with computational fluid dynamics modeling, provide the basis for redesign/reconfigurations as apropos. CFD is a very powerful and versatile tool for an analysis of this type.

At present, computational fluid dynamics methods are finding many new and diverse applications in bioengineering and biomimetics. For example, CFD techniques can be used to predict (1) velocity and stress distribution maps in complex reactor performance studies as well as in vascular and bronchial models; (2) strength of adhesion and dynamics of detachment for mammalian cells; (3) transport properties for nonhomogeneous materials and nonideal interfaces; (4) multicomponent diffusion rates using the Maxwell–Stefan transport model, as opposed to the limited traditional Fickian approach, incorporating interactive molecular immobilizing sites, and (5) materials processing capabilities useful in encapsulation technology and designing functional surfaces.

Although a full description of CFD techniques is beyond the scope of this chapter, thorough descriptions of the methods and procedures may be found in many texts, for example, Refs. [37–39].

References

1. Bird RB, Stewart WE, Lightfoot EN, 2002. *Transport Phenomena*, 2nd edn. Wiley, New York.
2. Lightfoot EN, 1974. *Transport Phenomena and Living Systems*, Wiley-Interscience, New York.
3. Cooney DO, 1976. *Biomedical Engineering Principles*, Dekker, New York.
4. Lightfoot EN, Duca KA, 2000. The roles of mass transfer in tissue function. In: Bronzino JD (ed) *The Biomedical Engineering Handbook*, 2nd edn. CRC Press, Boca Raton, Chapter 115.
5. Goldstein AS, DiMilla, PA, 1997. Application of fluid mechanic and kinetic models to characterize mammalian cell detachment in a radial-flow chamber, *Biotech. Bioeng.* 55:616.
6. Lauffenburger DA, Linderman JJ, 1993. *Receptors: Models for Binding, Trafficking, and Signaling*, Oxford University Press, New York.
7. Baldyga J, Bourne JR, 1999. *Turbulent Mixing and Chemical Reactions*, John Wiley and Sons, Ltd, Chicester, England.
8. Johnson B, Prud'homme R, 2003. Chemical processing and micro-mixing in confined impinging jets. *AIChE J.* 49(9):2264–82.
9. Lewis AS, Colton CK, 2004. Tissue engineering for insulin replacement in diabetes. In: Ma PX and Elisseeff J (eds.) *Scaffolding in Tissue Engineering*, Marcel Dekker, New York.
10. Schwarzer HC, Peukert W, 2004. Tailoring particle size through nanoparticle precipitation, *Chem. Eng. Comm.* 191:580–608.
11. Rabinow B, 2004. Nanosuspensions in drug delivery, *Nature Reviews-Drug Discovery* 3:785–96.
12. Rabinow B, 2005. Pharmacokinetics of drugs administered in nanosuspensions, *Discovery Medicine* 5(25):74–79.
13. Brennen CE, 2005. *Fundamentals of Multiphase Flow*, Cambridge University Press, NY.
14. Panagiotou T, Fisher RJ, 2008. Form nanoparticles via controlled crystallization: A bottom-up approach, *Chem. Eng. Prog.* 10(Oct.):33–39.
15. Panagiotou T, Mesite S, Fisher RJ, 2009. Production of norfloxacin nano-suspensions using microfluidics reaction technology (MRT) through solvent/anti solvent crystallization, *Ind. Eng. Chem. Res.* 48(4):1761–71.
16. Johnson AE, Fisher RJ, Weir GC, Colton CK, 2009. Oxygen consumption and diffusion in assemblages of respiring spheres: Performance enhancement of a bioartificial pancreas, *Chem. Eng. Sci.* 64(22):4470–87.
17. Gradl J, Peukert W, 2009. Simultaneous 3-D observation of different kinetic sub-processes for precipitation in a T-mixer, *Chem. Eng. Sci.* 64:709–20.
18. Fung YC, 1997. *Biomechanics: Circulation*, Springer-Verlag, New York.
19. Lamb H, 1945. *Hydrodynamics*. Dover Publishing, Inc., New York.
20. Schlichting H, 1979. *Boundary Layer Theory*, 7th edn. McGraw-Hill, New York.
21. Hinze JO, 1986. *Turbulence*, (Reissued). McGraw-Hill, New York.
22. Tennekes H, Lumley JL, 1972. *A First Course in Turbulence*, MIT Press, Cambridge.
23. Poiseuille JLM, 1840. Recherches experimentales sur le mouvement des liquids dans les tubes de tres petits diametres; 1. Influence de la pression sur la quantite de liquide qui traverse les tubes de tres petits diametres, *Comptes Rendus* 11:961.
24. Fox RW, McDonald AT, 1992. *Introduction to Fluid Mechanics*, 4th edn. John Wiley and Sons, New York.
25. Roberson JA, Crowe CT, 1997. *Engineering Fluid Mechanics*, 6th edn. John Wiley and Sons, New York.

26. Womersley JR, 1955. Method for the calculation of velocity, rate of flow and viscous drag in arteries when the pressure gradient is known, *J. Physiol.* 127:553.

27. Dwight HB, 1961. *Tables of Integrals and Other Mathematical Data*, McMillan Publishing Co., New York.

28. Gerrard JH, 1971. An experimental investigation of pulsating turbulent water flow in a tube, *J. Fluid. Mech.* 46(1):43.

29. Zamir M, 2000. *The Physics of Pulsatile Flow*, AIP Press, Springer-Verlag, New York.

30. Womersley JR, 1955. Oscillatory motion of a viscous liquid in a thin-walled elastic tube—I: The linear approximation for long waves, *Phil. Mag.* 46:199.

31. Atabek SC, Lew HS, 1966. An experimental investigation of pulsating turbulent water flow in a tube, *Biophys. J.* 6:481.

32. Eckmann DM, Grotberg JB, 1991. Experiments on transition to turbulence in oscillatory pipe flow, *J. Fluid Mech.* 222:329.

33. Davis SH, von Kerczek C, 1973. A reformulation of energy stability theory, *Arch. Rat. Mech. Anal.* 52:112.

34. von Kerczek C, Davis SH, 1974. Linear stability theory of oscillatory Stokes layers, *J. Fluid Mech.* 62:753.

35. Rahmani RK, Keith TG, Ayasoufi A, 2006. Numerical simulation and mixing study of pseudoplastic fluids in an industrial helical static mixer, *J. Fluids Eng.* 128:467.

36. Kumar V, Shirke V, Nigam KDP, 2008. Performance of Kenics static mixer over a wide range of Reynolds number, *Chem. Eng. J.* 139:284.

37. Fletcher CA, 1991. *Computational Techniques for Fluid Dynamics*, Volume I, 2nd edn. Springer-Verlag, Berlin.

38. Fletcher CA, 1991. *Computational Techniques for Fluid Dynamics*, Volume II, 2nd edn. Springer-Verlag, Berlin.

39. Chung TJ, 2002. *Computational Fluid Dynamics*, Cambridge University Press, Cambridge.

5

Animal Surrogate Systems

Michael L. Shuler
Cornell University

Sarina G. Harris
Cornell University

Xinran Li
Cornell University

Mandy B. Esch
Cornell University

5.1 Background..5-1
    Limitations of Animal Studies • Alternatives to Animal Studies
5.2 Cell Culture Analog Concept 5-2
5.3 Prototype CCAs...5-3
5.4 Models of Barrier Tissues and Their Use with μCCAs...............5-5
5.5 Future Prospects 5-8
Defining Terms 5-8
References..5-8

5.1 Background

Animal surrogate or cell culture analog (CCA) systems mimic the biochemical response of an animal or human when challenged with a chemical or drug. A true animal surrogate is a device that replicates the circulation, metabolism, and absorption of a chemical and its metabolites using interconnected multiple compartments to represent key organs. These compartments make use of engineered tissues or cell cultures. Physiologically based pharmacokinetic models (PBPK) guide the design of the device. The animal surrogate, particularly a human surrogate, can provide important insights into toxicity and efficacy of a drug or chemical when it is impractical or imprudent to use living animals (or humans) for testing. The combination of a CCA and PBPK provides a rational basis to relate molecular mechanisms to whole-animal response.

5.1.1 Limitations of Animal Studies

The primary method used to test the potential toxicity of a chemical or action of a pharmaceutical is to use animal studies, predominantly with rodents. However, animal studies are problematic. The primary difficulties are that the results may not be meaningful to the assessment of the human response (Gura, 1997). Because of the intrinsic complexity of a living organism and the inherent variability within a species, animal studies are difficult to use to identify unambiguously the underlying molecular mechanism for action of a chemical. The lack of a clear relationship among all of the molecular mechanisms to whole-animal response makes extrapolation across species difficult. This factor is particularly crucial when extrapolation of rodent data to humans is an objective. Further, without a good mechanistic model, it is difficult to rationally extrapolate from high doses to low doses. However, this disadvantage due to complexity can be an advantage; the animal is a "black box" and provides response data even when the mechanism of action is unknown. Further disadvantages reside in the high cost of animal studies, the long period of time often necessary to secure results, and the potential ethical problems in animal studies.

5.1.2 Alternatives to Animal Studies

In vitro methods using isolated cells (e.g., Del Raso, 1993) are inexpensive, quick, and have almost no ethical constraints (except the use of human embryonic stem cells). Because the culture environment can be specified and controlled, the use of isolated cells facilitates interpretation in terms of a biochemical mechanism. Since human cells can be used as well as animal cells, cross-species extrapolation is facilitated.

However, these techniques are not fully representative of human or animal response. Typical *in vitro* experiments expose isolated cells to a static dose of a chemical or drug. It is difficult to relate this static exposure to specific doses in a whole animal. The time-dependent change in the concentration of a chemical in an animal's organ cannot be replicated. If one organ modifies a chemical or prodrug that acts elsewhere, these actions would not be revealed by normal *in vitro* tests. Another related approach is the use of isolated cell cultures in a flow system such as a microphysiometer (McConnell et al., 1992; Cooke and O'Kennedy, 1999). Cells are cultured in a microscale (2.8 μL) flow cell, and changes in pH, measured electronically, report changes in cell physiology. An important use of this technology is the analysis or response of membrane-bound receptors in mammalian cells.

A major limitation of the use of cell cultures is that isolated cells do not fully represent the full range of biochemical activity of the corresponding cell type in a whole animal. Engineered tissues, especially cocultures (Bhatia et al., 1998), can provide a more authentic environment that can improve cell function. Another alternative is the use of tissue slices, typically from the liver (Olinga et al., 1997). However, tissue slices require the sacrifice of the animal, there is intrinsic variability, and biochemical activities can decay rapidly after harvest. The use of isolated tissue slices also does not reproduce interchange of metabolites among organs and the time-dependent exposure that occurs within an animal.

An alternative to both animal and *in vitro* studies is the use of computer models based on PBPK models (Connolly and Andersen, 1991). PBPK models can be applied to both humans and animals. Because PBPK models mimic the integrated, multicompartment nature of animals, they can predict the time-dependent changes in blood and tissue concentrations of a parent chemical or its metabolites. Although construction of a robust, comprehensive PBPK is time-consuming, once the PBPK is in place, many scenarios concerning exposure to a chemical or treatment strategies with a drug can be tested quickly and inexpensively. Since PBPKs can be constructed for both animals and humans, cross-species extrapolation is facilitated. There are, however, significant limitations in relying solely on PBPK models. PBPK models can only provide a response based on assumed mechanisms, and secondary and unexpected effects are not included. A further limitation is the difficulty in estimating parameters, particularly kinetic parameters.

None of these alternatives to animal studies predict the human response to chemicals or drugs satisfactorily.

5.2 Cell Culture Analog Concept

A CCA is a physical replica of the structure of a PBPK where cells or engineered tissues are used in organ compartments to achieve the metabolic and biochemical characteristics of the animal. The cell culture medium circulates between compartments and acts as a "blood surrogate." Small-scale bioreactors with the appropriate cell types in the physical device represent organs or tissues.

The CCA concept combines attributes of a PBPK and other *in vitro* systems, but unlike other *in vitro* systems, the CCA is an integrated system that can mimic dose dynamics and allows for conversion of a parent compound into metabolites and the interchange of metabolites between compartments. Because volume ratios of organs and compound residence times are replicated physiologically correctly, CCA systems allow for dose exposure scenarios that can replicate the exposure scenarios used in animal studies.

A CCA is intended to work in conjunction with a PBPK as a tool to test and refine mechanistic hypotheses. The PBPK can be made an exact replica of the CCA; the predicted response and measured CCA response should exactly match if the PBPK contains a complete and accurate description of the molecular mechanisms. In the CCA, all flow rates, the number of cells in each compartment, and the levels of each enzyme can be measured independently, so no adjustable parameters are required. If the PBPK predictions and CCA results disagree, then the description of the molecular mechanisms is incomplete. The CCA and PBPK can be used in an iterative manner to test modifications in the proposed mechanism. When the PBPK is extended to describe the whole animal, failure to predict animal response would be due to inaccurate description of transport (particularly within an organ), inability to accurately measure kinetic parameters such as *in vivo* enzyme levels or activities, or the presence of *in vivo* metabolic activities that are not present in the cultured cells or tissues.

The goal is predicting human pharmacological response to drugs or assessing risk due to chemical exposure. A PBPK that can make an accurate prediction of both animal CCA and animal experiments would be "validated." If we use the same approach to construct a human PBPK and CCA for the same compound, then we would have a rational basis to extrapolate animal response to predict human response when human experiments would be inappropriate. Further, since the PBPK is mechanistically based, it would provide a basis for extrapolation to low doses. The CCA/PBPK approach complements animal studies by potentially providing an improved basis for extrapolation to humans.

CCAs can also simulate synergistic or antagonistic behaviors of drugs or chemicals. If a PBPK for compound A and a PBPK for compound B are combined, then the response to any mixture of A and B should be predictable since the mechanisms for response to both A and B are included. Since CCAs are relatively inexpensive, many combinations of compounds A and B with different concentrations can be tested. Synergistic and antagonistic behaviors would be apparent from the behaviors of the cells and tissues cultured within the CCA device.

CCAs used in combination with PBPKs and validated with animal models provide a basis for predicting the human response to mixtures of drugs or chemicals.

5.3 Prototype CCAs

A simple three-component CCA mimicking rodent response to a challenge by naphthalene was developed and tested by Sweeney et al. (1995). While this prototype system did not fulfill the criteria for a CCA of physically realistic organ residence times or ratio of cell numbers in each organ, it did contain multiple compartments and was operated with fluid recirculation, which is necessary to capture the effects of metabolites. Rat hepatoma (H4IIE) cells and lung (L2) cells were used for the liver and the lung compartment, respectively. No cells were required in the "other tissue" compartment in this model since no metabolic reactions were postulated to occur elsewhere for naphthalene or its metabolites. The H4IIE cells contained enzyme systems for activation of naphthalene (cytochrome P450IA1) to the epoxide form and conversion to dihydriol (epoxide hydrolase) and conjugation with glutathione (glutathione-S-transferase). The L2 cells had no enzymes for naphthalene activation. Cells were cultured in glass vessels as monolayers. Experiments with this system using lactate dihydrogenase (LDH) release and glutathione levels as dependent parameters supported a hypothesis where naphthalene is activated in the "liver" and reactive metabolites circulate to the "lung," causing glutathione depletion and cell death as measured by LDH release. Increasing the level of cytochrome p450 activity in the "liver" by increasing cell numbers or by preinducing H4IIE cells led to increased death of L2 cells. Experiments with "liver"–blank, "lung"–"lung," and "lung"–blank combinations all supported the hypothesis of a circulating reactive metabolite as the cause of L2 cell death.

The prototype system (Sweeney et al., 1995) was difficult to operate, nonphysiologic, and made time course experiments difficult. An alternative system using packed bed reactors for the "liver" and "lung" compartments that was easier to operate was therefore developed and tested (Ghanem and Shuler, 2000a).

This system successfully allowed time course studies and was physiological with respect to the ratio of "liver" to "lung" cells. While liquid residence times improved in this system, they were still not physiologic (i.e., 114 s vs. an *in vivo* value of 21 s in the liver and 6.6 s vs. an *in vivo* lung value of about 1.5 s) due to physically limited flow through the packed beds. Unlike the prototype system, no response to naphthalene was observed.

This difference in response of the two CCA designs was explained through the use of PBPK models of each CCA (Ghanem and Shuler, 2000b). In the prototype system, the large liquid residence times in the liver and the lung allowed the formation of large amounts of naphthol from naphthalene oxide and presumably the conversion of naphthol into quinones that were toxic. In the packed bed system, liquid residence times were sufficiently small so that the predicted naphthol level was negligible. Thus, the PBPK provided a mechanistic basis to explain the differences in response of the two experimental configurations.

Using another simple CCA, Mufti and Shuler (1998) demonstrated that the response of human hepatoma cells (HepG2) to exposure to dioxin (2,3,7,8-tetrachlorodibenzo-*p*-dioxin) is dependent on how the dose is delivered. The induction of cytochrome p450IA1 activity was used as a model response for exposure to dioxin. Data were evaluated to estimate dioxin levels giving cytochrome P450IA1 activity 0.01% of maximal induced activity. Such an analysis mimics the type of analysis used to estimate the risk due to chemical exposure. The "allowable" dioxin concentration was 4×10^{-3} nM using a batch spinner flask, 4×10^{-4} nM using a one-compartment system with continuous feed, and 1×10^{-5} nM using a simple two-compartment CCA. Further, the response could be correlated to an estimate of the amount of dioxin bound to the cytosolic Ah receptor with a simple model for two different human hepatoma cell lines. This work illustrates the potential usefulness of a CCA approach in risk assessment.

Ma et al. (1997) have discussed an *in vitro* human placenta model for drug testing. This was a two-compartment perfusion system using human trophoblast cells attached to a chemically modified polyethylene therephthalate fibrous matrix as a cell culture scaffold. This system is a CCA in the same sense as the two-compartment system used to estimate the response to dioxin.

Integration of cell culture and microfabrication to form CCA or CCA-like systems has advanced rapidly in the last 10 years. The use of microfabricated devices allows for relatively high-throughput studies that are inexpensive, conserve scarce reagents and tissues, and facilitate automated collection and processing of data. The construction of a simple microscale CCA with multiple cell types and recirculating flow has been accomplished by Sin et al. (2004). The three-compartment system ("liver"–"lung"–other tissue) uses monolayer cultures of HepG2-C3A cells in the "liver" compartment and L2 cells in the "lung" compartment. While monolayer cultures are a poor representation of the physiology of real tissues, this system demonstrates that an "animal-on-a-chip" model is possible. A dissolved oxygen sensor using a fluorescent ruthenium complex was integrated into the system, demonstrating the potential to build real-time sensors into such a device.

The use of a microscale CCA for studying the toxicity of environmental contaminant has been demonstrated using naphthalene as a model toxicant. A silicon-based, microfabricated CCA with four chambers ("liver"–"lung"–"fat"–other tissue) was used for two studies (Viravaidya et al. 2004a, Viravaidya and Shuler, 2004b). These studies demonstrated that naphthalene is converted in the liver by P450IA1 into a reactive metabolite that circulates to the lung compartment. Further, the experiments show that 1,2-naphthalenediol and 1,2-naphthoquinone are the primary reactive metabolites that cause reduction in glutathione levels and cell death in the lung. Excess levels of 1-naphthol are converted to 1,2-naphthalenediol, a result that is consistent with those obtained in a prior study using the macroscale packed bed CCA (Ghanem and Shuler, 2000b). Naphthaquinone and naphthalenediol can be intraconverted through redox cycling generating reactive oxygen species. Naphthaquinone addition is toxic by itself. The addition of a fat compartment modulates the toxicity, providing significant, but partial, protection. These studies, together, demonstrate the utility of microscale CCAs for the simulation of the toxicity of toxicants present in the environment.

More recent studies with microscale CCAs have demonstrated their capability to capture the effects of cancer drugs. Using a 3-D µCCA device in which liver and colon cells were encapsulated in matrigel, the cytotoxic effects of Tegafur, a cancer drug that metabolizes in the liver to 5-fluorouracil (5-FU), was tested. The metabolite 5-FU acts as a chemotherapeutic agent for colon cancer. Operating the device without liver cells, Tegafur itself was effectively nontoxic to colon cancer cells (HCT-116). Adding liver cells (HepG2/C3A) to the system caused Tegafur to be converted to 5-FU by cytochrome P450 enzymes. The drug now exerted a significant toxic effect on HCT-116 cells (Sung and Shuler, 2009). This level of toxicity on HCT-116 cells was neither observed in 96-well plate experiments (with colon cells only) nor in µCCA experiments in which liver cells were absent. Results observed *in vitro* have previously only been seen in animal experiments or clinical studies involving humans. They confirm that µCCA devices are able to reproduce part of the liver metabolism and its consequences on HCT-116 cells.

Tatosian and Shuler (2009) first demonstrated that microscale CCAs could be used to simulate the synergistic effects of cancer drugs. In addition to the two uterine cancer cell lines MES-SA and its MDR variant MES-SA/DX-5, the device used for this study contained HepG2/C3A, representing the liver and metabolism of drugs, a megakaryoblast cell line (MEG-01), representing cells responsible for platelet formation, and a "normal" tissue compartment. Experiments were conducted with doxorubicin as the chemotherapeutic and two MDR suppressors, cyclosporine and nicardipine. Cyclosporine is used clinically as an immune system suppressor and nicardipine is a β-channel blocker. When either cyclosporine or nicardipine was added in addition to doxorubicin, the proliferation of the MDR cells was reduced from treatment with doxorubicin alone. More strikingly, when a combination of nicardipine and cyclosporine was used in place of a higher concentration of either MDR modulator alone, the MDR cell growth rate became negative. This synergistic interaction of the two modulators was not observed when using multiwell plate assays.

Examples of other devices with multicompartments that attempt to emulate aspects of human physiology include Chao et al. (2009), Vozzi et al. (2009), and Zhang et al. (2009). Chao et al. (2009) used a µCCA-type device with primary human liver cells to predict hepatic clearance of model drug compounds. Vozzi et al. (2009) used a related system to probe the interaction of murine hepatocytes with human vasculature endothelial cells (HUVECs) showing enhanced albumin and urea synthesis due to coculture within the system. Zhang et al. (2009) constructed a µCCA with four different cell types with local release of growth factors within a single compartment.

Simplifying the operation of µCCAs is an important step toward the development and use of multiorgan devices. Sung et al. (2010) employed a novel, multilayer design, which enhanced the usability of the devices while allowing hydrogel-cell cultures of multiple types. Gravity-induced flow enabled pumpless operation and prevented bubble formation. Three cell lines representing the liver, tumor tissue, and marrow were cultured in a three-chamber µCCA that was used to test the toxicity of the anticancer drug, 5-FU. The result was analyzed with a PK–PD model of the device, and compared with the result obtained in static cell culture. Each cell type exhibited differential responses to 5-FU, and the responses in the microfluidic environment were different from those in the static environment, but similar to what was anticipated from animal studies.

The above examples illustrate successful attempts to mimic the metabolic response of animals to drugs and environmental toxicants in integrated systems. The development of future CCAs will greatly benefit from engineered tissues that capture the authentic behavior of cells.

5.4 Models of Barrier Tissues and Their Use with µCCAs

While the number of drug leads is increasing, the capacity to increase animal and human clinical studies is limited. It is imperative that preclinical testing and predictions for human response become more accurate. CCAs could become important tools contributing to this end. Because barrier tissues can alter the physical and chemical properties of drugs as well as significantly influence a drug's bioavailability,

CCAs of tissues such as the skin, the gastrointestinal tract epithelium, and the lung epithelium are useful additions to the already existing models.

While only a few μCCAs that contain barrier tissue compartments have been developed so far, in principle, any *in vitro* model of a barrier tissue that was previously developed for drug testing with conventional methods could be adapted for use with μCCAs.

One of the first reports on the use of engineered cells by Gay et al. (1992) describes the use of a living skin equivalent as an *in vitro* dermatotoxicity model. The living skin equivalent consists of a coculture of human dermal fibroblasts in a collagen-containing matrix overlaid with human keratinocytes that have formed a stratified epidermis. Mitochondrial function was used to assess the toxicity of 18 different chemicals. Eleven compounds classified as nonirritating had minimal or no effect on mitochondrial activity. For seven known human skin irritants, the concentration that inhibited mitochondrial activity by 50% corresponded to the threshold value for each of these compounds to cause irritation on human skin. However, living skin equivalents did not fully mimic the barrier properties of human skin. For example, the permeability of water was 30-fold greater in the living skin equivalent than in human skin. In a study by Kriwet and Parenteau (1996) the permeabilities of 20 different compounds in *in vitro* skin models was reported. Comparisons indicate that skin cultures are slightly more permeable (two- or three-fold) for highly lipophilic substances and considerably more permeable (about 10-fold) for polar substances than human-cadaver or freshly excised human skin.

Validation of four *in vitro* tests for skin corrosion by the European Center for the Validation of Alternative Methods (ECVAM) has led to a combination of *in vitro* tests becoming mandatory for determining skin corrosion of chemicals in the European Union (Fentem and Botham 2002). These *in vitro* tests included a combination of rat skin electrical resistance measurements and commercial reconstituted skin equivalents (EpiDerm™ and EpiSkin™). After a series of prevalidation studies, the protocols have been further improved, resulting in the development of the EpiSkin model that has been validated as the stand-alone method of distinguishing irritants from nonirritants according to EU standards (Katoh et al. 2010). These developed skin equivalents are reconstituted human epidermal models. A skin model based on a cell line would be cheaper and more readily available. Suhonen et al. (2003) assessed a stratified rat epidermal keratinocyte cell line grown on a collagen gel at an air–liquid interface by measuring the permeability coefficients of 18 test compounds across the cell layer. The permeabilities were on average twofold greater than for human cadaver epidermis (range 0.3–5.2-fold difference). This cell culture model tended to overpredict the permeability of lipophilic solutes.

So far, the only μCCA that includes a skin compartment is that developed by Brand et al. (2000). Dorsal skin from male hairless mice was used in a perfusion system that contained a chamber with Hep-G2 liver cells in a compartment downstream of the skin compartment. The system was operated with a syringe pump and subjected to paroxovanadium [$VO(O_2)_2$ 1, 10 phenanthroline] bpV(phen). A net 22% increase in glucose consumption was measured in the Hep-G2 cells, demonstrating that the system is capable of simulating the uptake of the compound through skin. The authors also show that the system can be used with Caco-2 cell to construct a model of the intestinal epithelium (Brand et al. 2000).

One of the most used cell-based assays is the Caco-2 cell model of the intestine. This model can be used to determine the oral availability of a drug or chemical. Caco-2 cell cultures are derived from a human colon adenocarcinoma cell line. Artursson et al. (2001) reviewed the use of the Caco-2 cell line for the prediction of drug permeability and concluded that Caco-2 monolayers best predict the permeabilities of drugs that exhibit passive transcellular transport. For drug molecules transported by carrier proteins, the expression of the specific transport system in the Caco-2 monolayer needs to be characterized. The cell line, C2Bbel, is a clonal isolate of Caco-2 cells that is more homogeneous in apical brush border expression than the Caco-2 cell line. These cells form a polarized monolayer with an apical brush border morphologically comparable to the human colon. Tight junctions around the cells act to restrict passive diffusion by the paracellular route mimicking the transport resistance in the intestine. Hydrophobic solutes pass primarily by the transcellular route and hydrophilic compounds by the paracellular route. Yu and Sinko (1997) have demonstrated that the substratum (e.g., membrane) properties upon which the monolayer forms can become important in estimating the barrier properties of such *in vitro* systems. The barrier effects of the substratum need to be separated from the intrinsic property of the monolayers. Further, Anderle et al. (1998) have shown that the chemical nature of substratum and other culture conditions can alter transport properties. Sattler et al. (1997) provide one example (with hypericin) of how this model system can be used to evaluate effects of formulation (e.g., use of cyclodextrin or liposomes) on oral bioavailability. Another example is the application of the Caco-2 system to transport of paclitaxel across the intestine (Walle and Walle, 1998). Rapid passive transport was partially counterbalanced by an efflux pump (probably P-glycoprotein) limiting oral bioavailability.

Models that simulate first-pass metabolism combine Caco-2 cells with hepatic cells in transwells in which Caco-2 cells are cultured on porous membranes and HepG2/C3A cells are cultured in the chamber beneath. With such systems, the two-organ response can be partially recreated. For example, Caco-2 cells transport the toxin benzo[*a*]pyrene (B[*a*]P) and its metabolites back to the apical side, thereby preventing liver cells toxicity (Choi H et al. 2004). Thus, the known low bioavailability of the B[*a*]P was replicated *in vitro*. To reduce nutrient depletion over the course of 48-h experiments, a simple fluidic circuit that connects the two tissue compartments with each other can be constructed. Using such a system, the synergistic two-organ response to a challenge with 3-methylcholanthrene (3-MC) could be simulated (Choi SH et al. 2004). The induced activity of the enzyme CYP1A1/2 was more elevated than would have been expected from the individual cell cultures.

Mahler et al. have incorporated a Caco-2/MTX-HT29 coculture model into a CCA that contained several other organ compartments. When operated in the presence of liver cells, the uptake and metabolism of acetaminophen could be successfully simulated (Mahler et al., 2009). Both epithelial cells and liver cells metabolize acetaminophen, resulting in a dose-dependent decrease in liver cell viability. The results were within the range of those generated by a study of acetaminophen digestion in mice (Gujral et al. 2002).

Another barrier of interest for drug delivery studies is the blood–brain barrier (BBB). The BBB is formed by the endothelial cells of brain capillaries. The primary characteristics of the BBB are its high resistance to chemical diffusion and transport due to the presence of complex tight junctions that inhibit paracellular transport and its low endocytic activity. Several *in vitro* models of the BBB have been developed, and several authors have reviewed the models and their possible uses as permeability and toxicity screens (Reinhardt and Gloor, 1997; Gumbleton and Audus, 2001; Lundquist and Renftel, 2002). The most common *in vitro* BBB model consists of a monolayer of primary isolated brain capillary endothelial cells, primary isolated endothelial cells from elsewhere in the body, or an endothelial cell line cultured on a membrane insert. The endothelial cells are often cocultured with astrocytes or astroglial cells. In cocultures, the barrier properties of the BBB model increase.

The biggest challenge with *in vitro* BBB models is obtaining endothelial cell cultures that display extensive tight junctions as observed *in vivo*. According to de Boer et al. (1999), the large number of *in vitro* models and the accompanying diversity in laboratory techniques makes quantitative comparisons between models difficult. An example of an *in vitro* BBB system applied to a toxicological study is described by Glynn and Yazdanian (1998) who used bovine brain microvessel endothelial cells grown on porous polycarbonate filters to compare the transport of nevirapine, a reverse transcriptase inhibitor to other HIV antiretroviral agents. Nevirapine was the most permeable antiretroviral agent and hence may have value in HIV treatment in reducing levels of HIV in the brain.

The model developed by Stanness with endothelial cells and astrocytes cocultured on opposite sides of "capillaries" in a hollow-fiber reactor incorporates continuous physiological perfusion of the endothelial cells (Stanness et al., 1996). Harris and Shuler present a unique membrane, an order of magnitude thinner than those available commercially, for close contact coculture of endothelial and astrocytes (Harris and Shuler 2003; Harris Ma, 2004). The microfabricated membrane allows for the integration of the BBB model with other CCA compartments within microfluidic platforms.

Other barrier tissues that research has focused on are the bronchial and the eye epithelium. An epithelial/fibroblast coculture model of the bronchial epithelium was used to examine ozone toxicity (Lang et al. 1998). Huh et al. have used a microfluidic system to simulate the bronchial epithelium and were able to simulate and acoustically detect cellular-level lung injury induced by fluid mechanical stresses (Huh et al. 2007). Pasternak and Miller (1996) have tested a system to predict eye irritation, combining perfusion and a tissue model consisting of MDCK (Madin–Darby canine kidney) epithelial cells cultured on a semiporous cellulose ester membrane filter. The system could be fully automated using measurement of transepithelial electrical resistance (TER) as an end point. A decrease in TER is an indictor of cell damage and damage of the barrier function of the cell layer. The system was tested using nonionic surfactants and predicted the relative ocular toxicity of these compounds. The perfusion system mimics some dose scenarios (e.g., tearing) more easily than a static system and provides a more consistent environment for the cultured cells. A major advantage is that the TER can be measured throughout the entire exposure protocol without physically disturbing the tissue model and introducing artifacts in the response.

CCAs based on the concepts described here and incorporating advanced-engineered models of barrier tissues could become a powerful tool for testing the bioavailability of pharmaceuticals.

5.5 Future Prospects

The most serious bottleneck in pharmaceutical development is the ability to complete ADMET (adsorption–distribution–metabolism–elimination–toxicity) studies early enough in the development process to focus resources on the best drug candidates. Of particular importance are human surrogates that can improve the probability that a drug will be successful in clinical trials. Such trials may cost more than a 100 million dollars and success at the rate of one in three rather than current values (about one in ten) would offer significant economic advantage.

Over the last 4 years, the development of integrated devices that combine cell culture and microfabrication makes the commercial applications to pharmaceutical evaluation a real possibility (see Freedman, 2004, for discussion). However, the authenticity of engineered tissues remains a hurdle. While tissue with low levels of vascularization (e.g., skin and cartilage) can be mimicked reasonably well, vascularized tissues (e.g., liver) are still quite challenging. As improvements in tissue engineering occur, one of the first applications will be in testing of chemicals and pharmaceuticals. Over the next 10 years, we expect CCA-type systems to become industrially important in preclinical testing of pharmaceuticals and in evaluating chemicals (and chemical mixtures) for toxicity.

Defining Terms

Animal surrogate: A physiologically based cell or tissue multicompartmented device with fluid circulation to mimic metabolism and fate of a drug or chemical.
Engineered tissues: Cell culture mimic of a tissue or organ; often combines a polymer scaffold and one or more cell types.
Physiologically based pharmacokinetic model (PBPK): A computer model that replicates animal physiology by subdividing the body into a number of anatomical compartments, each compartment interconnected through the body fluid systems; used to describe the time-dependent distribution and disposition of a substance.
Tissue slice: A living organ is sliced into thin sections for use in toxicity studies; one primary organ can provide material for many tests.

References

Anderle P, Niederer E, Werner R, Hilgendorf C, Spahn-Langguth H, Wunderu-Allenspach H, Merkle HP, Langguth P. 1998. P-Glycoprotein (P-gp) mediated efflux in Caco-2 cell monolayers: The influence of culturing conditions and drug exposure on P-gp expression levels. *J. Pharm. Sci.* 87:757.

Artursson P, Palm K, Luthman K. 2001. Caco-2 monolayers in experimental and theoretical predictions of drug transport. *Adv. Drug Del. Rev.* 46:27.

Bhatia SN, Balis UJ, Yarmush ML, Toner M. 1998. Microfabrication of hepatocyte/fibroblast co-cultures: Role of homotypic cell interactions. *Biotechnol. Prog.* 14:378.

Brand RM, Hannah TL, Mueller C, Cetin Y, Hamel FG. 2000. A novel system to study the impact of epithelial barriers on cellular metabolism. *Ann. Biomed. Eng.* 28:1210.

Chao P, Maguire T, Novik E. 2009. Evaluation of a microfluidic based cell culture platform with primary human hepatocytes for the predictin of hepatic clearance in human. *Biochem. Pharmacol.* 78(6):625.

Choi H, Nishikawa M, Sakoda A, Sakai, Y. 2004. Feasibility of a simple double-layered coculture system incorporating metabolic processes of the intestine and liver tissue: Application to the analysis of benzo[*a*]pyrene toxicity. *Toxicol. In Vitro* 18:393.

Choi SH, Fukuda O, Sakoda A, Sakai Y. 2004. Enhanced cytochrome P450 capacities of Caco-2 and Hep G2 cells in new coculture system under the static and perfused conditions: Evidence for possible organ-to-organ interactions against exogenous stimuli. *Mater. Sci. Eng. C* 24:333.

Connolly RB, Andersen ME. 1991. Biologically based pharmacodynamic models: Tool for toxicological research and risk assessment. *Annu. Rev. Pharmacol. Toxicol.* 31:503.

Cooke D, O'Kennedy R. 1999. Comparison of the detrazolium salt assay for succinate dehydrogenase with the cytosensor microphysiometer in the assessment of compound toxicities. *Anal. Biochem.* 274:188–194.

de Boer AG, Gaillard PJ, Breimer, DD. 1999. The transference of results between blood-brain barrier cell culture systems. *Eur. J. Pharm. Sci.* 8:1.

Del Raso NJ. 1993. *In vitro* methodologies for enhanced toxicity testing. *Toxicol. Lett.* 68:91

Fentem JH, Botham PA. 2002. ECVAM's activities in validating alternative tests for skin corrosion and irritation. *Altern. Lab. Anim.* 30(Suppl 2):61.

Freedman, DH. 2004. The silicon guinea pig. *Technol. Rev.* 107(June):62.

Gay R, Swiderek M, Nelson D, Ernesti A. 1992. The living skin equivalent as a model *in vitro* for ranking the toxic potential of dermal irritants. *Toxic. In Vitro* 6:303.

Ghanem A, Shuler, ML. 2000a. Characterization of a perfusion reactor utilizing mammalian cells on microcarrier beads. *Biotechnol. Prog.* 16:471–479.

Ghanem A, Shuler ML. 2000b. Combining cell culture analogue reactor designs and PBPK models to probe mechanisms of naphthalene toxicity. *Biotechnol. Prog.* 16:334.

Glynn SL, Yazdanian Y. 1998. *In vitro* blood-brain barrier permeability of nevirapine compared to other HIV antiretroviral agents. *J. Pharm. Sci.* 87:306.

Gujral JS, Knight TR, Farhood A, Bajt ML, Jaeschke H. 2002. Mode of cell death after acetaminophen overdose in mice: Apoptosis or oncotic necrosis? *Toxicol. Sci.* 67(2):322.

Gumbleton M, Audus KL. 2001. Progress and limitations in the use of *in vitro* cell cultures to serve as a permeability screen for the blood-brain barrier. *J. Pharm. Sci.* 90:1681.

Gura T. 1997. Systems for identifying new drugs are often faulty. *Science* 273:1041.

Harris Ma S. 2004. A physiologically based *in vitro* model of the blood-brain barrier utilizing a nanofabricated membrane. PhD Thesis. Cornell University, Ithaca, New York.

Harris S, Shuler ML. 2003. Growth of endothelial cells on microfabricated silicon nitride membranes for an *in vitro* model of the blood-brain barrier. *Biotechnol. Bioprocess Eng.* 8:246.

Huh D, Fujioka H, Tung Y, Futai N, Paine III R, Grotberg JB, Takayama S. 2007. Acoustically detectable cellular-level lung injury induced by fluid mechanical stresses in microfluidic airway systems. *PNAS* 104:18886.

Katoh M, Hamajima F, Ogasawara T, Hata K. 2010. Assessment of human epidermal model LabCyte EPI-MODEL for *in vitro* skin irritation testing according to European Centre for the Validation of Alternative Methods (ECVAM)-validated protocol. *J. Toxicol. Sci.* 34(3):327.

Kriwet K, Parenteau NL. 1996. *In vitro* skin models. *Cosmetics Toiletries* 111(Feb):93.

Lang DS, Jorres RA, Mucke M, Siegfried W, Magnussen H. 1998. Interactions between human bronchoepithelial cells and lung fibroblasts after ozone exposure in vitro. *Toxicol. Lett.* 96,97:13.

Lundquist S, Renftel M. 2002. The use of *in vitro* cell culture models for mechanistic studies and as permeability screens for the blood-brain barrier in the pharmaceutical industry—Background and current status in the drug discovery process. *Vasc. Pharmacol.* 38:335.

Ma T, Yang S-T, Kniss DA. 1997. Development of an *in vitro* human placenta model by the cultivation of human trophoblasts in a fiber-based bioreactor system. *Am. Inst. Chem. Eng. Ann. Mtg.*, Los Angeles, CA, Nov. 16–21.

Mahler GJ, Esch MB, Glahn RP, Shuler ML. 2009. Characterization of a gastrointestinal tract microscale cell culture analog used to predict drug toxicity. *Biotechnol Bioeng.* 104(1):193.

McConnell HM, Owicki JC, Parce JW, Miller DL, Baxter GT, Wada HG, Pitchford S. 1992. The cytometer microphysiometer: Biological applications of silicon technology. *Science* 257:1906.

Mufti NA, Shuler ML. 1998. Different *in vitro* systems affect CYPIA1 activity in response to 2,3,7,8-tetrachlorodibenzo-*p*-dioxin. *Toxicol. In Vitro* 12:259.

Olinga, P, Meijer DKF, Slooff, MJH, Groothuis, GMM. 1997. Liver slices in *in vitro* pharmacotoxicology with special reference to the use of human liver tissue. *Toxicol. In Vitro* 12:77.

Pasternak AS, Miller WM. 1996. Measurement of trans-epitheial electrical resistance in perfusion: Potential application for *in vitro* ocular toxicity testing. *Biotechnol. Bioeng.* 50:568.

Reinhardt CA, Gloor SM. 1997. Co-culture blood-brain barrier models and their use for pharmatoxicological screening. *Toxicol. In Vitro.* 11:513.

Sattler S, Schaefer U, Schneider W, Hoelzl J, Lehr C-M. 1997. Binding, uptake, and transport of hypericin by Caco-2 cell monolayers. *J. Pharm. Sci.* 86:1120.

Sin A, Chin KC, Jamil MF, Kostov Y, Rao G, Shuler ML. 2004. The design and fabrication of three-chamber microscale cell culture analog devices with integrated dissolved oxygen sensors. *Biotechnol. Prog.* 20:338.

Stanness KA, Guatteo E, Janigro D. 1996. A dynamic model of the blood-brain barrier "*in vitro.*" *NeuroToxicol.* 17:481.

Suhonen TM, Pasonen-Seppanen S, Kirjavainen M, Tammi M, Tammi R, Urtti A. 2003. Epidermal cell culture model derived from rat keratinocytes with permeability characteristics comparable to human cadaver skin. *Eur. J. Pharm. Sci.* 20:107.

Sung JH, Kam, C, Shuler ML. 2010. A microfluidic device for a pharmacokinetic–pharmacodynamic (PK–PD) model on a chip. *Lab Chip* 10:446.

Sung JH, Shuler ML 2009. A micro cell culture analog (microCCA) with 3-D hydrogel culture of multiple cell lines to assess metabolism-dependent cytotoxicity of anti-cancer drugs. *Lab Chip* 9:1385.

Sweeney LM, Shuler ML, Babish JG, Ghanem A. 1995. A cell culture analog of rodent physiology: Application to naphthalene toxicology. *Toxicol. In Vitro* 9:307.

Tatosian DA, Shuler ML. 2009. A novel system for evaluation of drug mixtures for potential efficacy in treating multidrug resistant cancers. *Biotechnol. Bioeng.* 103:187.

Walle UK, Walle T. 1998.Taxol transport by human intestinal epithelial Caco-2 cells. *Drug Metabol. Disposit.* 26:343.

Yu H, Sinko PJ. 1997. Influence of the microporous substratum and hydrodynamics on resistances to drug transport in cell culture systems: Calculation of intrinsic transport parameters. *J. Pharm. Sci.* 86:1448.

Viravaidya K, Sin A, Shuler ML. 2004a. Development of a microscale cell culture analog to probe naphthalene toxicity. *Biotechnol. Prog.* 20:316.

Viravaidya K, Shuler ML. 2004b. Incorporation of 3T3-L1 cells to mimic bioaccumulation in a microscale cell culture analog device for toxicity studies. *Biotechnol. Prog.* 20:590.

Vozzi F, Heinrich JM, Bader A, Ahluwali AD. 2009. Connected culture of murine hepatocytes and HUVEC in a multicompartmental bioreactor. *Tissue Eng Part A* 15(6):1291.

Zhang C, Zhao Z, Abdul Rahim NA. van Noort D, Yu H. 2009. Towards a human-on-chip: Culturing multiple cell types on a chip with compartmentalized microenvironments. *Lab Chip.* 9(22):3185.

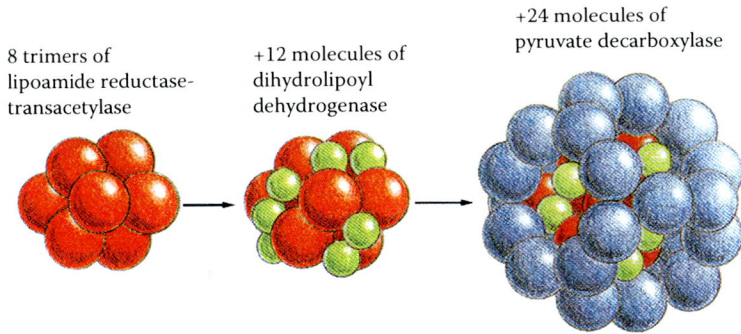

FIGURE 2.4 The pyruvate dehydrogenase complex.

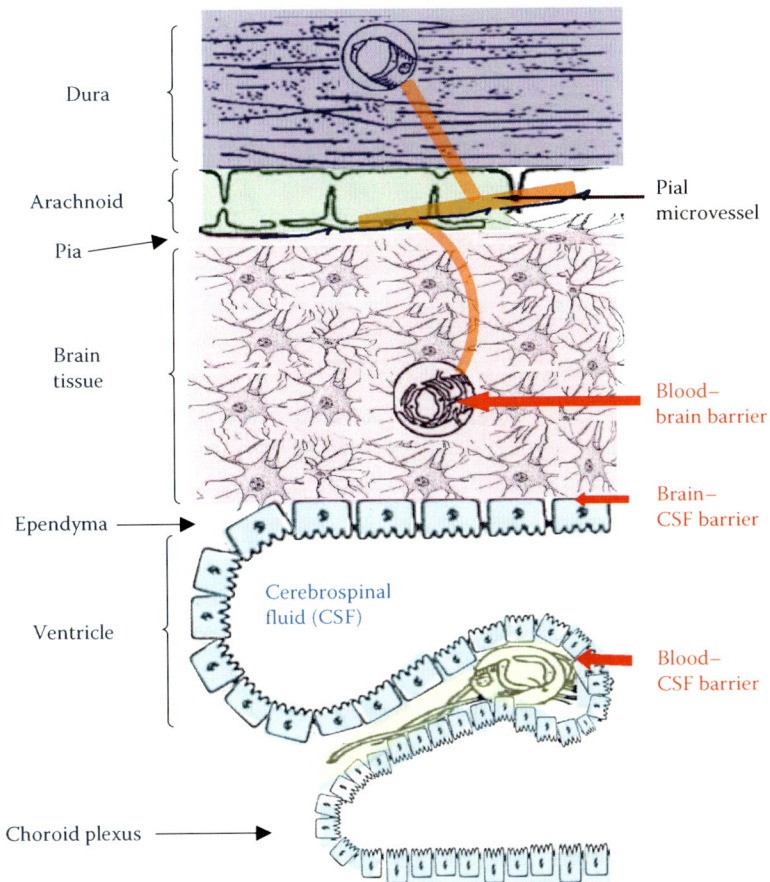

FIGURE 8.1 Locations of barriers in the brain. (Modified from Abbott NJ, 2004. *Neurosci Int.* 45:545.)

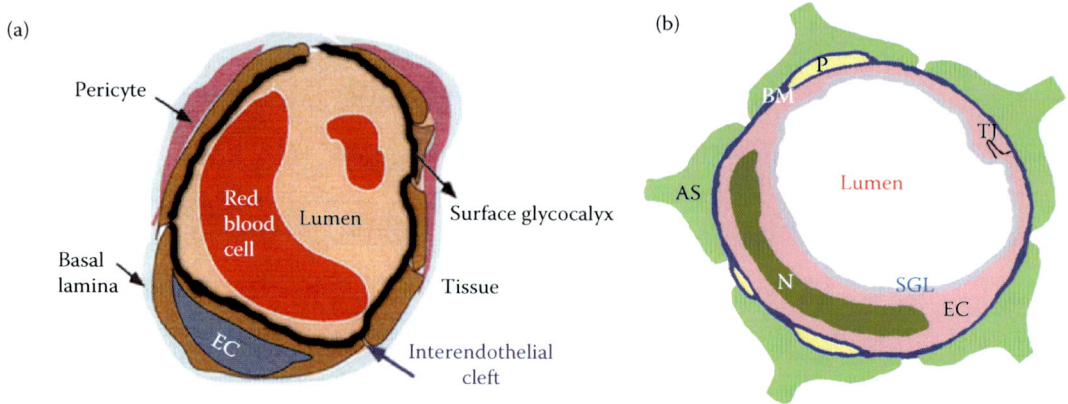

FIGURE 8.2 Schematic of the cross-sectional view of (a) a peripheral microvessel (the microvessel in nonbrain organs), and (b) the blood–brain barrier (BBB) or cerebral microvessel (the microvessel in the brain). In addition to other structures as in a peripheral microvessel, the BBB is wrapped by astrocyte foot processes (AS, green). BM, basement membrane (or basal lamina); EC, endothelial cell; N, nucleus of endothelial cell; P, pericytes; SGL, surface glycocalyx layer; TJ, tight junction.

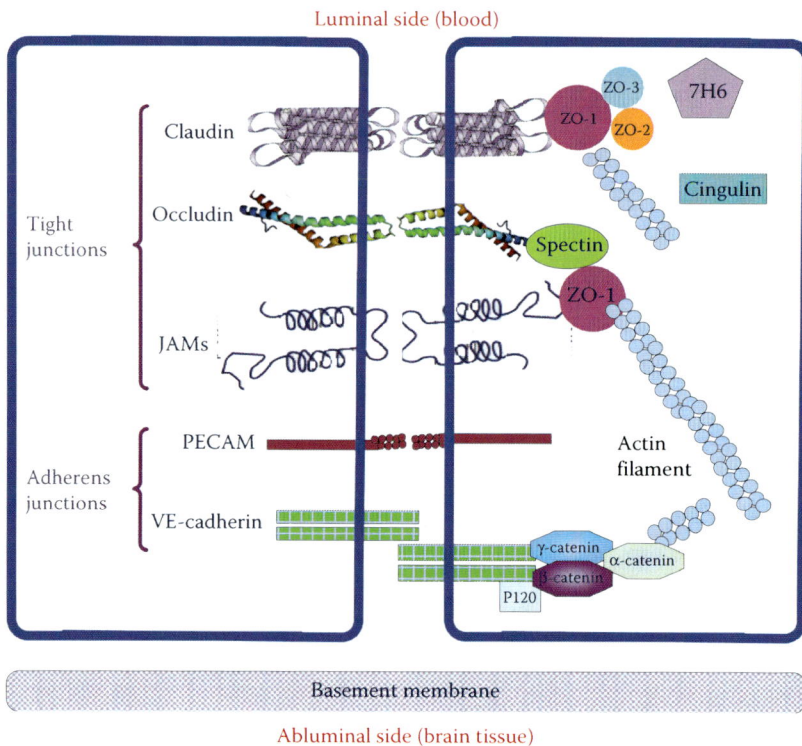

FIGURE 8.3 Schematic of junctional complex in the paracellular pathway of the BBB. (Modified from Kim JH et al. 2006. *J Biochem Mol Biol.* 39(4):339.)

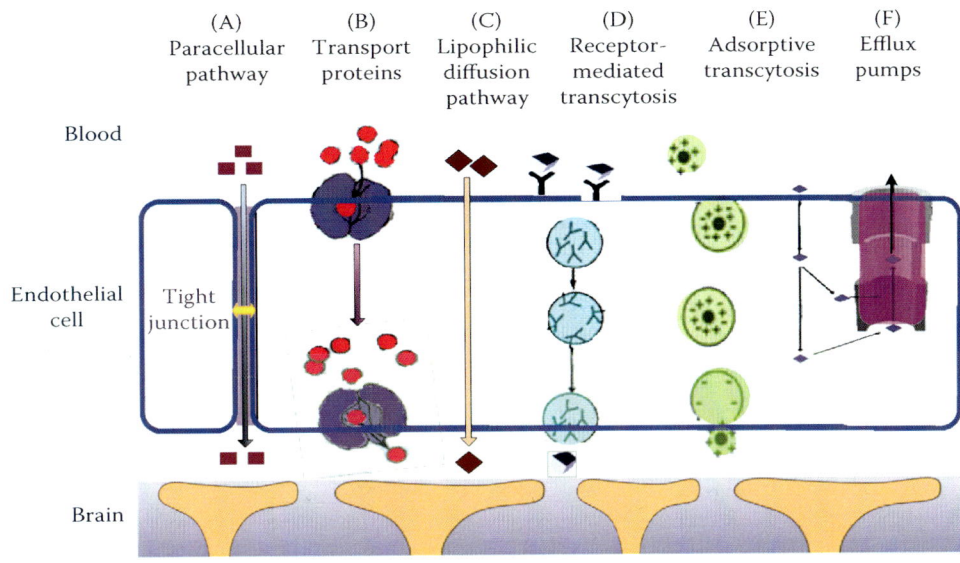

FIGURE 8.4 Transport pathways across the brain endothelial cell. (Modified from Neuwelt EA, 2004. *Neurosurgery* 54(1):131.)

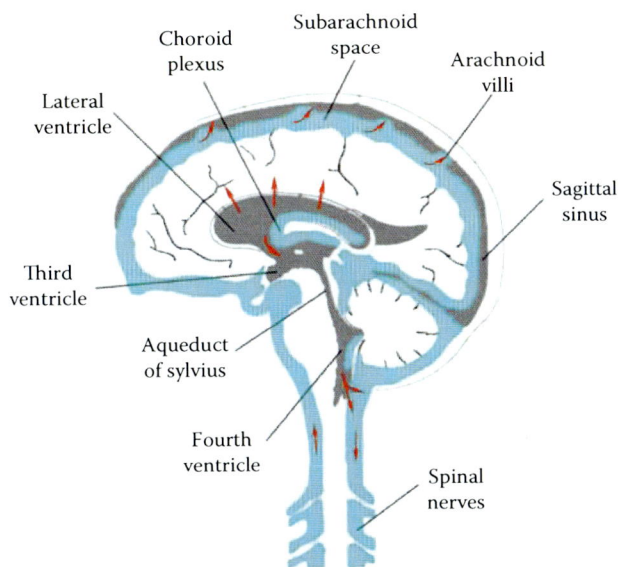

FIGURE 8.5 Circulation of the cerebrospinal fluid (CSF) and the brain–CSF barrier. (Modified from Abbott NJ et al. 2010. *Neurobiol Dis.* 37:13.)

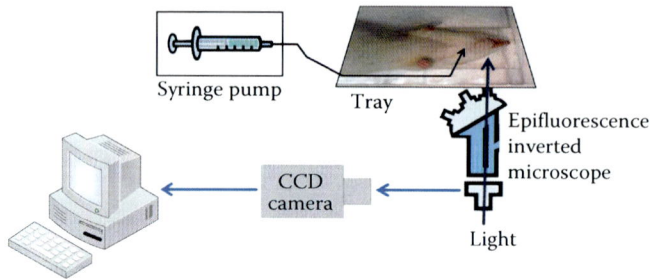

FIGURE 8.6 Schematic for the *in vivo* permeability measurement of rat pial microvessels. The fluorescence solution was injected into the brain via a carotid artery with a syringe pump. The fluorescence images were captured by a CCD camera, which was connected to an inverted microscope. The image analysis software was then used to measure the fluorescence intensity for the region of interest in each image.

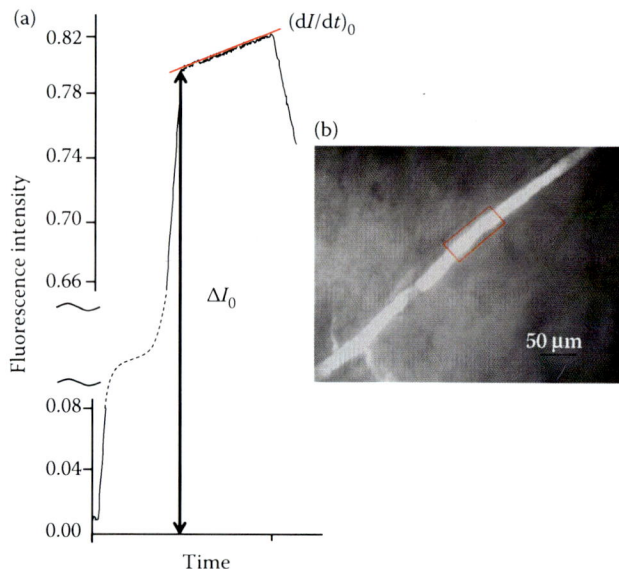

FIGURE 8.7 Quantitative fluorescence imaging method for the measurement of solute permeability in a rat pial microvessel. The images were collected during the *in vivo* experiments and the fluorescence intensity was analyzed off-line. When the fluorescence labeled test solute was injected into the carotid artery, the pial microvessel lumen filled with fluorescent solute (red frame in b), producing ΔI_0. With continued perfusion, the measured fluorescence intensity increased indicating further transport of the solute out of the microvessel and into the surrounding tissue. The initial solute flux into the tissue was measured from the slope $(dI/dt)_0$ (a). The solute permeability P was calculated by $P = 1/\Delta I_0 \ (dI/dt)_0 \ r/2$. Here, r is the microvessel radius. The scale bar in (b) is 50 μm. (Redrawn from Yuan W et al. 2009. *Microvasc Res.* 77:166.)

6

Arterial Wall Mass Transport: The Possible Role of Blood Phase Resistance in the Localization of Arterial Disease

6.1   Steady-State Transport Modeling 6-2
       Reactive Surface • Permeable Surface • Reactive Wall

6.2   Damköhler Numbers for Important Solutes 6-5
       Adenosine Triphosphate • Albumin and LDL • Oxygen

6.3   Sherwood Numbers in the Circulation 6-6
       Straight Vessels

6.4   Nonuniform Geometries Associated with Atherogenesis 6-7
       Sudden Expansion • Stenosis • Bifurcation • Curvature

6.5   Discussion 6-11

6.6   Possible Role of Blood Phase Transport in Atherogenesis 6-12
       Direct Mechanical Effects on Endothelial Cells • Hypoxic Effect on
       Endothelial Cells • Hypoxia Induces VEGF

References 6-13

John M. Tarbell
The City College of New York

Yuchen Qiu
Cordis Corporation

Atherosclerosis is a disease of the large arteries that involves a characteristic accumulation of high-molecular-weight lipoprotein in the arterial wall [1]. The disease tends to be localized in regions of curvature and branching in arteries where fluid shear stress (shear rate) is altered from its normal patterns in straight vessels [2]. The possible role of fluid mechanics in the localization of atherosclerosis has been debated for many years [3,4]. One possibility considered early on was that the blood phase resistance to lipid transport, which could be affected by local fluid mechanics, played a role in the focal accumulation of lipid in arteries. Studies by Caro and Nerem [5], however, showed that the uptake of lipid in arteries could not be correlated with fluid phase mass transport, leading to the conclusion that the wall (endothelium), and not the blood, was the limiting resistance to lipid transport. This suggested that fluid mechanical effects on macromolecular transport were the result of direct mechanical influences on the transport characteristics of the endothelium.

While the transport of large molecules such as low-density lipoprotein (LDL) and other high-molecular-weight materials, which are highly impeded by the endothelium, may be limited by the wall and not the fluid (blood), other low-molecular-weight species which undergo rapid reaction on the endothelial surface (e.g., adenosine triphosphate—ATP) or which are consumed rapidly by the underlying tissue (e.g., oxygen) may be limited by the fluid phase. With these possibilities in mind, the purpose of this short review is to compare the rates of transport in the blood phase to the rates of reaction on the endothelial surface, the rates of transport across the endothelium, and the rates of consumption within the wall of several important biomolecules. It will then be possible to assess quantitatively the importance of fluid phase transport; to determine which molecules are likely to be affected by local fluid mechanics; to determine where in blood vessels these influences are most likely to be manifest; and finally, to speculate about the role of fluid phase mass transport in the localization of atherosclerosis.

6.1 Steady-State Transport Modeling

6.1.1 Reactive Surface

Referring to Figure 6.1, we will assume that the species of interest is transported from the blood vessel lumen, where its bulk concentration is C_b, to the blood vessel surface, where its concentration is C_s, by a convective–diffusive mechanism that depends on the local fluid mechanics and can be characterized by a fluid-phase mass transfer coefficient k_L (see Reference 6 for further background). The species flux in the blood phase is given by

$$J_s = k_L(C_b - C_s) \tag{6.1}$$

At the endothelial surface, the species may undergo an enzyme-catalyzed surface reaction (e.g., the hydrolysis of ATP to ADP) which can be modeled using classical Michaelis–Menten kinetics with a rate given by

$$V = \frac{V_{max}C_s}{k_m + C_s} \tag{6.2}$$

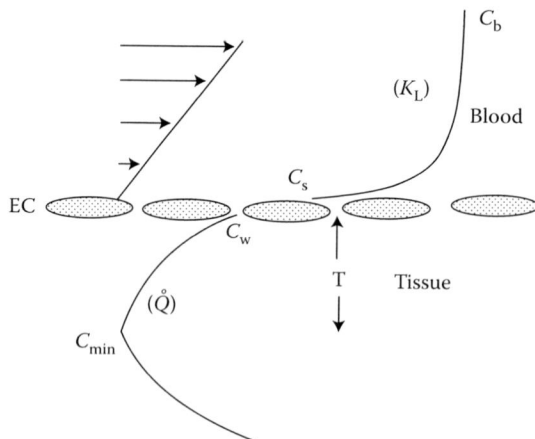

FIGURE 6.1 Schematic diagram of arterial wall transport processes showing the concentration profile of a solute which is being transported from the blood, where its bulk concentration is C_b, to the surface of the endothelium, where its concentration is C_s, then across the endothelium, where the subendothelial concentration is C_w, and finally to a minimum value within the tissue, C_{min}. Transport of the solute in the blood phase is characterized by the mass transport coefficient, k_L, consumption of the solute at the endothelial surface is described by a first-order reaction with rate constant, k_r; movement of the solute across the endothelium depends on the permeability coefficient, Pe; and reaction of the solute within the tissue volume is quantified by a zeroeth-order consumption rate, $\overset{\circ}{Q}$.

where V_{max} is the maximum rate (high C_s) and k_m is the Michaelis constant. When $C_s \ll k_m$, as is often the case, then the reaction rate is pseudo-first order

$$V = k_r C_s \qquad (6.3)$$

with the rate constant for the surface reaction given by $k_r = V_{max}/k_m$.

At steady state, the transport to the surface is balanced by the consumption at the surface so that

$$k_L(C_b - C_s) = k_r C_s \qquad (6.4)$$

It will be convenient to cast this equation into a dimensionless form by multiplying it by d/D, where d is the vessel diameter and D is the diffusion coefficient of the transported species in blood, or the media of interest. Equation 6.4 then becomes

$$Sh(C_b - C_s) = Da_r C_s \qquad (6.5)$$

where

$$Sh \equiv \frac{k_L d}{D} \qquad (6.6)$$

is the Sherwood number (dimensionless mass transfer coefficient), and

$$Da_r \equiv \frac{k_r d}{D} \qquad (6.7)$$

is the Damkhöler number (dimensionless reaction rate coefficient). Solving Equation 6.5 for the surface concentration, one finds

$$\frac{C_s}{C_b} = \frac{1}{1 + Da_r/Sh} \qquad (6.8)$$

When $Da_r \ll Sh$,

$$C_s = C_b \qquad (6.9)$$

and the process is termed "wall-limited" or "reaction-limited." On the other hand, when $Da_r \gg Sh$,

$$C_s = \left(\frac{Sh}{Da_r} \right) C_b \qquad (6.10)$$

and the process is termed "transport-limited" or "fluid phase-limited." It is, in this transport-limited case, that the surface concentration, and in turn the surface reaction rate, depends on the fluid mechanics which determines the Sherwood number. It will therefore be useful to compare the magnitudes of Da_r and Sh to determine whether fluid mechanics plays a role in the overall transport process of a surface reactive species.

6.1.2 Permeable Surface

Many species will permeate the endothelium without reacting at the luminal surface (e.g., albumin, LDL) and their rate of transport (flux) across the surface layer can be described by

$$J_s = Pe(C_s - C_w) \tag{6.11}$$

where Pe is the endothelial permeability coefficient and C_w is the wall concentration beneath the endothelium. If the resistance to transport offered by the endothelium is significant, then it will be reasonable to assume

$$C_w \ll C_s \tag{6.12}$$

so that at steady state when the fluid and surface fluxes balance

$$k_L(C_b - C_s) = PeC_s \tag{6.13}$$

Multiplying Equation 6.13 by d/D to introduce dimensionless parameters and then solving for the surface concentration leads to

$$\frac{C_s}{C_b} = \frac{1}{1 + Da_e/Sh} \tag{6.14}$$

where Sh is defined in Equation 6.6 and

$$Da_e \equiv \frac{Ped}{D} \tag{6.15}$$

is a Damkhöler number based on endothelial permeability. Equation 6.14 shows that when $Da_e \ll Sh$, the transport process is again "wall-limited." When $Da_e \gg Sh$, fluid mechanics again becomes important through the Sherwood number.

6.1.3 Reactive Wall

Oxygen is transported readily across the endothelium (Hellums), but unlike most proteins, is rapidly consumed by the underlying tissue. In this case, it is fair to neglect the endothelial transport resistance (assume $C_w = C_s$), and then by equating the rate of transport to the wall with the (zeroeth order) consumption rate within the wall we obtain

$$K_L(C_b - C_s) = \mathring{Q}T \tag{6.16}$$

where $\mathring{Q}$ is the tissue consumption rate and T is the tissue thickness (distance from the surface to the minimum tissue concentration—see Figure 6.1). For the specific case of O_2 transport, it is conventional to replace concentration (C) with partial pressure (P) through the Henry's law relationship $C = KP$, where K is the Henry's law constant. Invoking this relationship and rearranging Equation 6.16 into a convenient dimensionless form, we obtain

$$\frac{P_s}{P_b} = 1 - \frac{Da_w}{Sh} \tag{6.17}$$

where Sh is defined in Equation 6.6, and Da_w is another Damkhöler number based on the wall consumption rate

$$Da_w = \frac{\mathring{Q}Td}{KDP_b} \tag{6.18}$$

Clearly, when $Da_w \ll Sh$, the process is wall-limited. But, as $Da_w \rightarrow Sh$, the process becomes limited by transport in the fluid phase ($P_s \rightarrow 0$), and fluid mechanics plays a role. Because we are treating the tissue consumption rate as a zeroeth-order reaction, the case $Da_w > Sh$ is not meaningful ($P_s < 0$). In reality, as Sh is reduced, the tissue consumption rate must be reduced due to the lack of oxygen supply from the blood.

6.2 Damkhöler Numbers for Important Solutes

A wide range of Damkhöler numbers characterizes the transport of biomolecular solutes in vessel walls of the cardiovascular system, and in this section, we focus on four important species as examples of typical biotransport processes: ATP, a species that reacts vigorously on the endothelial surface, albumin, and LDL, species that are transported across a permeable endothelial surface; and oxygen, which is rapidly consumed within the vessel wall. Since most vascular disease (atherosclerosis) occurs in vessels between 3 and 10 mm in diameter, we use a vessel of 5 mm diameter to provide estimates of typical Damkhöler numbers.

6.2.1 Adenosine Triphosphate

ATP is degraded at the endothelial surface by enzymes (ectonucleotidases) to form adenosine diphosphate (ADP). The Michaelis–Menten kinetics for this reaction has been determined by Gordon et al. [7] using cultured porcine aortic endothelial cells: $k_m = 249$ μM, $V_{max} = 22$ nmol/min/10^6 cells. V_{max} can be converted to a molar flux by using a typical endothelial cell surface density of 1.2×10^5 cells/cm^2, with the result that the pseudo-first-order rate constant (Equation 6.3) is $k_r = 1.77 \times 10^{-4}$ cm/s. Assuming a diffusivity of 5.0×10^{-6} cm^2/s for ATP [8], and a vessel diameter of 5 mm, we find

$$Da_r = 17.7$$

6.2.2 Albumin and LDL

These macromolecules are transported across the endothelium by a variety of mechanisms including nonspecific and receptor-mediated trancytosis, and paracellular transport through normal or "leaky" interendothelial junctions [9,10]. In rabbit aortas, Truskey et al. [11] measured endothelial permeability to LDL and observed values on the order of Pe = 1.0×10^{-8} cm/s in uniformly permeable regions, but found that permeability increased significantly in punctate regions associated with cells in mitosis to a level of Pe = 5×10^{-7} cm/s. Using this range of values for Pe, assuming a diffusivity of 2.5×10^{-7} cm^2/s for LDL, and a vessel diameter of 5 mm, we find

$$Da_e = 0.02\text{--}1.0 \text{ (LDL)}$$

For albumin, Truskey et al. [12] reported values of the order Pe = 4.0×10^{-8} cm/s in the rabbit aorta. This presumably corresponded to regions of uniform permeability. They did not report values in punctate regions of elevated permeability. More recently, Lever et al. [13] reported Pe values of similar magnitude in the thoracic and abdominal aorta as well as the carotid and renal arteries of rabbits. In the ascending aorta and pulmonary artery, however, they observed elevated permeability to albumin on the order of Pe = 1.5×10^{-7} cm/s. Assuming a diffusivity of 7.3×10^{-7} cm²/s for albumin, a vessel diameter of 5 mm, and the range of Pe values described above, we obtain

$$Da_e = 0.027\text{–}0.10 \text{ (albumin)}$$

6.2.3 Oxygen

The first barrier encountered by oxygen after being transported from the blood is the endothelial layer. Although arterial endothelial cells consume oxygen [14], the pseudo-first-order rate constant for this consumption is estimated to be an order of magnitude lower than that of ATP, and it is therefore reasonable to neglect the endothelial cell consumption relative to the much more significant consumption by the underlying tissue. Liu et al. [15] measured the oxygen permeability of cultured bovine aortic and human umbilical vein endothelial cells and obtained values of 1.42×10^{-2} cm/s for bovine cell monolayers and 1.96×10^{-2} cm/s for human cell monolayers. Because the endothelial permeability to oxygen is so high, it is fair to neglect the transport resistance of the endothelium and to direct attention to the oxygen consumption rate within the tissue.

To evaluate the Damköhler number based on the tissue consumption rate (Equation 6.17), we turn to data of Buerk and Goldstick [16] for $\mathring{Q}/(KD)$ measured both *in vivo* and *in vitro* in dog, rabbit, and pig blood vessels. The values of $\mathring{Q}/(KD)$ reported by Buerk and Goldstick are based on tissue properties for KD. To translate these tissue values into blood values, as required in our estimates (Equation 6.17), we use the relationship $(KD)_{\text{tissue}} = N\,(KD)_{\text{water}}$ suggested by Paul et al. [17] and assume $(KD)_{\text{blood}} = (KD)_{\text{water}}$. In the thoracic aorta of dogs, $\mathring{Q}/(KD)$ ranged from 1.29×10^5 torr/cm² to 5.88×10^5 torr/cm² in the tissue. The thickness (distance to the minimum tissue O_2 concentration) of the thoracic aorta was 250 µm and the diameter is estimated to be 0.9 cm [18]. PO_2 measured in the blood (P_b) was 90 torr. Introducing these values into Equation 6.17, we find

$$Da_w = 10.8\text{–}49.0 \text{ (thoracic aorta)}$$

In the femoral artery of dogs, $\mathring{Q}/(KD)$ ranged from 35.2×10^5 torr/cm² to 46.9×10^5 torr/cm² in the tissue. The thickness of the femoral artery was 50 µm and the estimated diameter is 0.4 cm [18]. PO_2 measured in the blood was about 80 torr. These values lead to the following estimates:

$$Da_w = 29.3\text{–}39.1 \text{ (femoral artery)}$$

6.3 Sherwood Numbers in the Circulation

6.3.1 Straight Vessels

For smooth, cylindrical tubes (a model of straight blood vessels) with well-mixed entry flow, one can invoke the thin concentration boundary layer theory of Lévêque [6] to estimate the Sherwood number in the entry region of the vessel where the concentration boundary is developing. This leads to

$$Sh = 1.08 x^{*-1/3} \quad \text{(constant wall concentration)} \tag{6.19a}$$

$$1.30 x^{*-1/3} \quad \text{(constant wall flux)} \tag{6.19b}$$

where

$$x^* = \frac{x/d}{Re \cdot Sc} \tag{6.20}$$

TABLE 6.1 Transport Characteristics in a Straight Aorta

Species	Sc	x^*	Sh	Da
O_2	2900	4.1×10^{-5}	31.1	10.8–49.0
ATP	7000	1.7×10^{-5}	41.8	17.7
Albumin	48,000	2.5×10^{-6}	79.2	0.027–0.100
LDL	140,000	8.6×10^{-7}	114	0.02–1.00

Note: $d = 1$ cm, $x = 60$ cm, Re = 500, $v = .035$ cm²/s.

is a dimensionless axial distance which accounts for differing rates of concentration boundary layer growth due to convection and diffusion. In Equation 6.20, Re = vd/v is the Reynolds number, Sc = v/D is the Schmidt number, and their product is the Peclet number. Equation 6.19 is quite accurate for distances from the entrance satisfying $x^* < .001$. Sh continues to drop with increasing axial distance as the concentration boundary layer grows, as described by the classical Graetz solution of the analogous heat transfer problem [19]. When the concentration boundary layer becomes fully developed, Sh approaches its asymptotic minimum value

$$Sh = 3.66 \quad \text{(constant wall concentration)} \tag{6.21a}$$

$$Sh = 4.36 \quad \text{(constant wall flux)} \tag{6.21b}$$

For a straight vessel, Sh cannot drop below these asymptotic values. Equations 6.19 and 6.21 also indicate that the wall boundary condition has little effect on the Sherwood number.

It is instructive to estimate Sh at the end of a straight tube having dimensions and flow rate characteristics of the human aorta (actually a tapered tube). Table 6.1 compares Sh and Da (for O_2, ATP, albumin, and LDL) at the end of a 60-cm-long model aorta having a diameter of 1 cm and a flow characterized by Re = 500.

Table 6.1 clearly reveals that for a straight aorta, transport is in the entry or Lévêque regime ($x^* < 10^{-3}$). For albumin and LDL, Da ≪ Sh, and transport is expected to be "wall-limited." For O_2 and ATP, Da ~ Sh, and the possibility of "fluid phase-limited" transport exists. At the lowest possible rates of wall mass transport in a straight vessel (Sh = 3.66–4.36), transport is still expected to be "wall-limited" for albumin and LDL, whereas it would be "fluid-phase limited" for oxygen and ATP.

6.4 Nonuniform Geometries Associated with Atherogenesis

6.4.1 Sudden Expansion

Flow through a sudden expansion (Figure 6.2) at sufficiently high Reynolds number induces flow separation at the expansion point followed by reattachment downstream. This is a simple model of physiological flow separation. The separation zone is associated with low wall shear stress since this quantity is identically zero at the separation and reattachment points.

An experimental study of oxygen transport in saline and blood for an area expansion ratio of 6.7 and Reynolds numbers in the range 160–850 [20] displayed the general spatial distribution of Sh displayed in Figure 6.2. The minimum value of Sh was observed near the separation point and the maximum value appeared near the reattachment point. The maximum Sh ranged between 500 and several thousand depending on the conditions and the minimum value was ~50. A numerical study of the analogous heat transfer problem by Ma et al. [21] at a lower area expansion ratio (4.42) and Schmidt number (Sc = 105) showed the same qualitative trends indicated in Figure 6.2, but the Sherwood numbers were considerably lower (between 4 and 40) due to the lower expansion ratio, lower Schmidt number, and different entrance conditions. In both these studies, Sh did not drop below its fully developed tube flow value at any axial location.

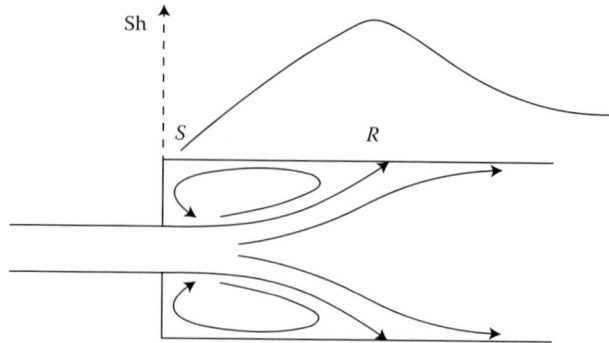

FIGURE 6.2 Schematic diagram showing the spatial distribution of the Sherwood number downstream of a sudden expansion. The flow separates (S) from the wall at the expansion point and reattaches (R) downstream. The Sherwood number is reduced near the separation point (radial velocity away from the wall) and elevated near the reattachment point (radial velocity toward the wall).

The sudden expansion flow field provides insight into the mechanism controlling the spatial distribution of the Sherwood number in separated flows. Near the reattachment point, the radial velocity component convects solute toward the wall (enhancing transport), whereas near the separation point, the radial velocity component convects solute away from the wall (diminishing transport). The net result of this radial convective transport superimposed on diffusive transport (toward the wall) is a maximum in Sh near the reattachment point and a minimum in Sh near the separation point.

6.4.2 Stenosis

Flow through a symmetric stenosis at sufficiently high Reynolds number (Figure 6.3) will lead to flow separation at a point downstream of the throat and reattachment further downstream. Again, because the wall shear stress is identically zero at the separation and reattachment points, the separation zone is a region of low wall shear stress. Conversely, converging flow upstream of the stenosis induces elevated wall shear stress.

Schneiderman et al. [22] performed numerical simulations of steady flow and transport in an axisymmetric, 89% area restriction stenosis with a sinusoidal axial wall contour at various Reynolds numbers for a Schmidt number typical of oxygen transport. Relative to a uniform tube, Sh was always elevated

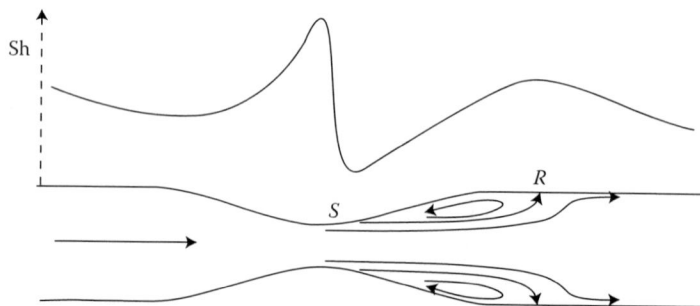

FIGURE 6.3 Schematic diagram showing the spatial distribution of the Sherwood number around a symmetric stenosis. The flow converges upstream of the stenosis where the Sherwood number is elevated (radial velocity toward the wall). The flow separates (S) from the wall just downstream of the throat, if the Reynolds number is high enough, and reattaches (R) downstream. The Sherwood number is reduced near the separation point (radial velocity away from the wall) and elevated near the reattachment point (radial velocity toward the wall).

in the converging flow region upstream of the stenosis, and when Re was low enough to suppress flow separation, transport remained elevated downstream of the stenosis as well. At high Re (Figure 6.3), flow separation produced a region of diminished transport (relative to a uniform tube) while reattachment induced elevated transport. The lowest value of Sh in the diminished transport regime was ~10. This occurred at Re = 16, a flow state for which separation was incipient. At high Re, the minimum Sh was elevated above 10, and the region of diminished transport was reduced in axial extent.

More recent numerical simulations by Rappitsch and Perktold [23,24] for a 75% stenosis in steady flow (Re = 448, Sc = 2000 (oxygen)) and sinusoidal flow (Re = 300, Sc = 46,000 (albumin)) show the same basic trends depicted in Figure 6.3. Again, Sh was reduced in a narrow region around the separation point, but did not drop below ~10.

Moore and Ethier [25] also simulated oxygen transport in a symmetric stenosis, but they accounted for the binding of oxygen to hemoglobin and solved for both the oxygen and oxyhemoglobin concentrations. They determined axial Sh profiles throughout the stenosis, which showed the same basic tendencies indicated in Figure 6.3. However, because hemoglobin has a much higher molecular weight and Schmidt number than oxygen (about 100 times higher Sc), the local values of Sh were higher than computed on the basis of free oxygen transport alone. This is expected since simple transport considerations (Lévêque solution) suggest $Sh \propto Sc^{1/3}$.

The Sh profile for the stenosis (Figure 6.3) reflects the same underlying mechanisms that were operative in the sudden expansion (Figure 6.2): radial flow directed toward the wall enhances transport near the reattachment point, while radial flow directed away from the wall diminishes transport near the separation point.

6.4.3 Bifurcation

A few numerical studies of flow and transport in the carotid artery bifurcation have been reported recently as summarized in Figure 6.4. The carotid artery bifurcation is a major site for the localization of atherosclerosis, predominantly on the outer wall (away from the flow divider) in the flow separation zone, which is a region of low and oscillating wall shear stress [26]. Perktold et al. [27] simulated O_2 transport in a realistic pulsatile flow through an anatomically realistic three-dimensional carotid bifurcation geometry using a constant wall concentration boundary condition. Ma et al. [28] simulated oxy-

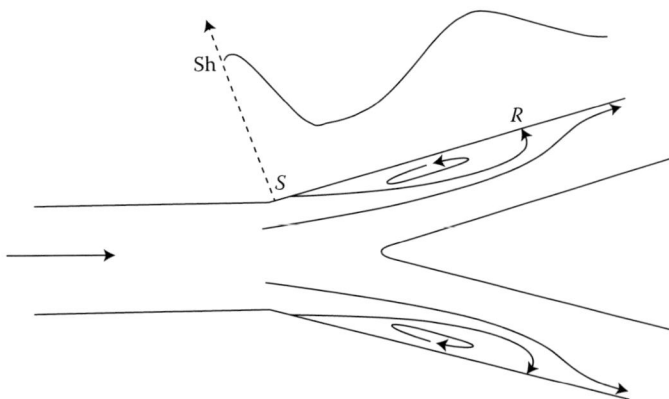

FIGURE 6.4 Schematic diagram showing the spatial distribution of the Sherwood number along the outer wall of a bifurcation. The flow separates (*S*) if the Reynolds number is high enough and there is an increase in cross-sectional area through the bifurcation, and reattaches (*R*) downstream. The Sherwood number is reduced near the separation point (radial velocity away from the wall) and elevated near the reattachment point (radial velocity toward the wall).

gen transport in steady flow through a realistic, three-dimensional, carotid bifurcation with a constant wall concentration boundary condition.

As in the sudden expansion and stenosis geometries, the bifurcation geometry can induce flow separation on the outer wall with reattachment downstream. Again there is a region of attenuated transport near the separation point and amplified transport near the reattachment point. Perktold et al. [27] predicted minimum Sherwood numbers close to zero in the flow separation zone. Ma et al. [28] predicted the same general spatial distribution, but the minimum Sherwood number was ~25. Differences in the minimum Sherwood number may be due to differing entry lengths upstream of the bifurcation as well as differences in flow pulsatility.

6.4.4 Curvature

Localization of atherosclerosis has also been associated with arterial curvature [2]. For example, the inner curvature of proximal coronary arteries as they bend over the curved surface of the heart has been associated with plaque localization [29]. Qiu [30] carried out three-dimensional, unsteady flow computations in an elastic (moving wall) model of a curved coronary artery for O_2 transport with a constant wall concentration boundary condition. He obtained the results shown in Figure 6.5. Because

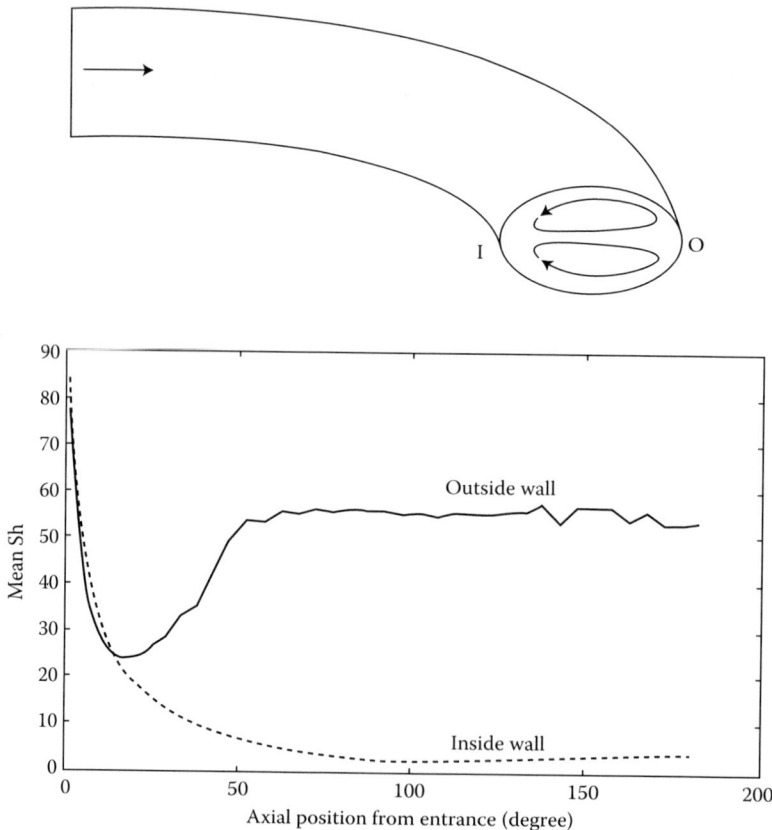

FIGURE 6.5 Schematic diagram showing the spatial distribution of the Sherwood number along the inner (I—toward the center of curvature) and outer (O—away from the center of curvature) walls of a curved vessel. In the entry region, before the secondary flow has developed, the Sherwood number follows a Lévêque distribution. As the secondary flow evolves, the Sherwood number becomes elevated on the outer wall where the radial velocity of the secondary flow is toward the wall, and diminished on the inner wall where radial velocity of the secondary flow is away from the wall.

Qiu assumed uniform axial velocity and concentration profiles (with no secondary flow) at the curved tube entrance (0 degrees), there is an entrance region (~25 degrees) where Sh essentially follows a Lévêque boundary layer development. Eventually (~50 degrees), the secondary flow effects become manifest, and marked differences in transport rates between the inside wall (low transport: Sh ~2) and the outside wall (high transport: Sh ~55) develop. The regions of high and low transport in the curved vessel geometry cannot be associated with axial flow separation (as in the expansion, stenosis, and bifurcation) because flow separation does not occur at the modest curvature levels in the coronary artery simulation.

The secondary flow (in the plane perpendicular to the axial flow—Figure 6.5) determines, in large measure, the differences in transport rates between the inside and outside walls. The radial velocity is directed toward the wall at the outside, leading to enhancement of transport by a convective mechanism. At the inside of the curvature, the radial velocity is directed away from the wall, and transport is impeded by the convective mechanism. This secondary flow mechanism produces transport rates that are 25 times higher on the outside wall than the inside wall. This is to be contrasted with wall shear stress values which, for the coronary artery condition, are <2 times higher on the outside wall than on the inside wall [31]. Earlier studies of fully developed, steady flow and transport in curved tubes [32] are consistent with the above observations.

6.5 Discussion

The considerations of mass transport in the fluid (blood) phase and consumption by the vessel wall described in the preceding sections indicate that only highly reactive species such as O_2 and ATP can be "transport-limited." Larger molecules such as albumin and LDL, which are not rapidly transformed within the vessel wall, are not likely to be transport-limited.

For O_2, ATP, and other molecules characterized by large values of the Damkhöler number, transport limitation will occur in regions of the circulation where the Sherwood number is low (Sh < Da). Localized regions of low Sh arise in nonuniform geometries around flow separation points (not reattachment points) where radial flow velocities are directed away from the wall, and in secondary flow regions where the secondary velocity is directed away from the wall (inside wall of a curved vessel).

Considerable supporting experimental evidence for the above conclusions is available, and a few representative studies will be mentioned here. Santilli et al. [33] measured transarterial wall oxygen tension gradients at the dog carotid bifurcation. Oxygen tensions at the carotid sinus (outer wall in Figure 6.4) were significantly decreased in the inner 40% of the artery wall compared to an upstream control location. Oxygen tensions at the flow divider (inner wall in Figure 6.4) were increased significantly throughout the artery wall compared with control locations. These observations are consistent with fluid phase-limited oxygen transport at the outer wall of the carotid bifurcation.

Dull et al. [8] measured the response of intracellular free calcium in cultured bovine aortic endothelial cells (BAECs) after step changes in flow rate, using media containing either ATP or a nonreactive analog of ATP. In the presence of ATP, which is rapidly degraded by enzymes on the endothelial cell surface, step changes in flow rate induced rapid changes in intracellular calcium which were not apparent when the inactive ATP analog was used in place of ATP. The interpretation of these experiments was that increases in flow increased the rate of mass transport of ATP to the cell surface and exceeded the capacity of the surface enzymes to degrade ATP. This allowed ATP to reach surface receptors and stimulate intracellular calcium. This study thus provided evidence that ATP transport to the endothelium could be fluid phase (transport)-limited.

Caro and Nerem [5] measured the uptake of labeled cholesterol bound to serum lipoprotein in excised dog arteries under well-defined conditions in which the Lévêque solution (Equation 6.19) described the fluid phase mass transport process. If the transport of lipoprotein to the surface had been controlled by the fluid phase, they should have observed a decrease in uptake with distance from the vessel entrance following a $x^{-1/3}$ law (Equation 6.19). They did not, however, observe any significant spatial variation of uptake of lipoprotein over the length of the blood vessel. This observation is consistent with "wall-limited" transport of lipoprotein as we have suggested in the preceding sections.

6.6 Possible Role of Blood Phase Transport in Atherogenesis

Accumulation of lipid in the arterial intima is a hallmark of atherosclerosis, a disease that tends to be localized on the outer walls of arterial bifurcations and the inner walls of arterial curvatures [2]. The outer walls of bifurcations and the inner walls of curvatures may have localized regions characterized by relatively low blood phase transport rates (low Sh in Figures 6.4 and 6.5). But, how can low transport rates lead to high accumulation of lipid in the wall? If, as we have argued in this review, LDL transport is really not affected by the blood phase fluid mechanics, but is limited by the endothelium, how can local fluid mechanics influence intimal lipid accumulation? There are several possible scenarios for a fluid mechanical influence which are reviewed briefly.

6.6.1 Direct Mechanical Effects on Endothelial Cells

The outer walls of bifurcations and the inner walls of curved vessels are characterized by low mean wall shear stress, and significant temporal oscillations in wall shear stress direction (oscillatory shear stress) [2]. Endothelial cells in this low, oscillatory shear environment tend to assume a polyhedral, cobblestone enface morphology, whereas endothelial cells in high shear regions tend to be elongated in the direction of flow. It has been suggested that these altered morphologies, which represent chronic adaptive responses to altered fluid mechanical environments, are characterized by distinct macro-molecular permeability characteristics [2,10] as direct responses of the endothelial layer to altered mechanical environments. In addition, a number of studies have shown that fluid shear stress on the endothelial surface can have an acute influence on endothelial transport properties both *in vitro* [34–37] and *in vivo* [38,39].

6.6.2 Hypoxic Effect on Endothelial Cells

Hypoxia (low oxygen tension), which can be induced by a blood phase transport limitation, can lead to a breakdown of the endothelial transport barrier either by a direct effect on the endothelial layer or by an indirect mechanism in which hypoxia upregulates the production of hyperpermeabilizing cytokines from other cells in the arterial wall. A number of recent studies have shown that hypoxia increases mac-romolecular transport across endothelial monolayers in culture due to metabolic stress [40–42]. These studies describe direct effects on the endothelial layer since other cells present in the vessel wall were not present in the cell culture systems.

6.6.3 Hypoxia Induces VEGF

Many cell lines express increased amounts of vascular endothelial growth factor (VEGF) when sub-jected to hypoxic conditions, as do normal tissues exposed to hypoxia, functional anemia, or localized ischemia [43]. VEGF is a multifunctional cytokine that acts as an important regulator of angiogenesis and as a potent vascular permeabilizing agent [43,44]. VEGF is believed to play an important role in the hyperpermeability of microvessels in tumors, the leakage of proteins in diabetic retinopathy, and other vascular pathologies [45,46]. Thus, a plausible scenario for the increase in lipid uptake in regions of poor blood phase mass transport is the following: Hypoxia upregulates the production of VEGF by cells within the vascular wall and the VEGF in turn permeabilizes the endothelium, allowing increased transport of lipid into the wall. This mechanism can be depicted schematically as follows:

$$O_2 \downarrow \rightarrow \text{VEGF} \uparrow \rightarrow \text{Pe} \uparrow$$

In support of this view, several recent studies have shown that VEGF is enriched in human atherosclerotic lesions [47,48]. Smooth muscle cells and macrophages appear to be the predominant sources of VEGF in such lesions. Thus, a mechanism in which hypoxia induces VEGF and hyperpermeability is plausible, but at the present time it must only be considered a hypothesis relating fluid phase transport limitation and enhanced macromolecular permeability.

References

1. Ross, R., Atherosclerosis: A defense mechanism gone awry, *Am. J. Pathol.*, 143, 987, 1993.
2. Nerem, R., Atherosclerosis and the role of wall shear stress, in: *Flow-Dependent Regulation of Vascular Function*, Bevan, J. A., Kaley, G., and Rubany, G. M., Eds., Oxford University Press, New York, 1995.
3. Caro, L. G., Fitz-Gerald, J. M., and Schroter, R. C., Atheroma and arterial wall shear: Observation, correlation and proposal of a shear dependent mass transfer mechanism for atherogenesis, *Proc. R. Soc. London (Biol.)*, 177, 109, 1971.
4. Fry, D. L., Acute vascular endothelial changes associated with increased blood velocity gradients, *Circ. Res.*, 22, 165, 1968.
5. Caro, C. G. and Nerem, R. M., Transport of 14C-4-cholesterol between serum and wall in the perfused dog common carotid artery, *Circ. Res.*, 32, 187, 1973.
6. Basmadjian, D., The effect of flow and mass transport in thrombogenesis, *Ann. Biomed. Eng.*, 18, 685, 1990.
7. Gordon, E. L., Pearson, J. D., and Slakey, L. L., The hydrolysis of extra cellular adenine nucleotides by cultured endothelial cells from pig aorta, *J. Biol. Chem.*, 261, 15496, 1986.
8. Dull, R. O., Tarbell, J. M., and Davies, P. F., Mechanisms of flow-mediated signal transduction in endothelial cells: Kinetics of ATP surface concentrations, *J. Vasc. Res.*, 29, 410, 1992.
9. Lin, S. J., Jan, K. M., Schuessler, G., Weinbaum, S., and Chien, S., Enhanced macromolecular permeability of aortic endothelial cells in association with mitosis, *Atherosclerosis*, 17, 71, 1988.
10. Weinbaum, S. and Chien, S., Lipid transport aspects of atherogenesis, *J. Biomech. Eng.*, 115, 602, 1993.
11. Truskey, G. A., Roberts, W. L., Herrmann, R. A., and Malinauskas, R. A., Measurement of endothelial permeability to 125I-low density lipoproteins in rabbit arteries by use of en face preparations, *Circ. Res.*, 71, 883, 1992.
12. Truskey, G. A., Colton, C. K., and Smith, K. A., Quantitative analysis of protein transport in the arterial wall, in: *Structure and Function of the Circulation*, 3, Schwartz, C. J., Werthessen, N. T., and Wolf, S., Eds., Plenum Publishing Corp., New York, NY, 1981, p. 287.
13. Lever, M. J., Jay, M. T., and Coleman, P. J., Plasma protein entry and retention in the vascular wall: Possible factors in atherogenesis, *Can. J. Physiol. Pharmacol.*, 74, 818, 1996.
14. Motterlini, R., Kerger, H., Green, C. J., Winslow, R. M., and Intaglietta, M., Depression of endothelial and smooth muscle cell oxygen consumption by endotoxin. *Am. J. Physiol.*, 275, H776, 1998.
15. Liu, C. Y., Eskin, S. G., and Hellums, J. D., The oxygen permeability of cultured endothelial cell monolayers, paper presented at the *20th International Society of Oxygen Transport to Tissue Conference*, August 26–30, Mianz, Germany, 1992.
16. Buerk, D. G. and Goldstick, T. K., Arterial wall oxygen consumption rate varies spatially, *Am. J. Physiol.*, 243, H948, 1982.
17. Paul, R. J., Chemical energetics of vascular smooth muscle, in: *Handbook of Physiology. The Cardiovascular System. Vascular Smooth Muscle*, Vol. 2, Sheperd, J. T. and Abboud, F. M., Eds., Am. Physiol. Soc., Bethesda, MD, 1980, p. 201.
18. Caro, C. G., Pedley, T. J., Schroter, R. C., and Seed, W. A., *The Mechanics of the Circulation*, Oxford University Press, Oxford, 1978.
19. Bennett, C. O. and Meyers, J. E., *Momentum, Heat and Mass Transfer*, McGraw-Hill, New York, 1962.
20. Thum, T. F. and Diller, T. E., Mass transfer in recirculating blood flow, *Chem. Eng. Commun.*, 47, 93, 1986.

21. Ma, P., Li, X., and Ku, D. N., Heat and mass transfer in a separated flow region of high Prandtl and Schmidt numbers under pulsatile conditions, *Int. J. Heat Mass Transfer*, 37(17), 2723–2736, 1994.
22. Schneiderman, G., Ellis, C. G., and Goldstick, T. K., Mass transport to walls of stenosed arteries: Variation with Reynolds number and blood flow separation, *J. Biomech.*, 12, 869, 1979.
23. Rappitsch, G. and Perktold, K., Computer simulation of convective diffusion processes in large arteries, *J. Biomech.*, 29, 207, 1996.
24. Rappitsch, G. and Perktold, K., Pulsatile albumin transport in large arteries: A numerical simulation study, *J. Biomech. Eng.*, 118, 511, 1996.
25. Moore, J. A. and Ethier, C. R., Oxygen mass transfer calculations in large arteries, *J. Biomech. Eng.*, 119, 469, 1997.
26. Ku, D. N., Giddens, D. P., Zarins, C. K., and Glagov, S., Pulsatile flow and atherosclerosis in the human carotid bifurcation: Positive correlation between plaque location and low and oscillating shear stress, *Arteriosclerosis*, 5, 293, 1985.
27. Perktold, K., Rappitsch, G., Hofer, M., and Karner, G., Numerical simulation of mass transfer in a realistic carotid artery bifurcation model, *Proceedings of the 1997 Bioengineering Conference (ASME)*, BED, 35, 85, 1997.
28. Ma, P., Li, X., and Ku, D. N., Convective mass transfer at the carotid bifurcation, *J. Biomech.*, 30, 565, 1997.
29. Chang, L. J. and Tarbell, J. M., A numerical study of flow in curved tubes simulating coronary arteries, *J. Biomech.*, 21, 927, 1988.
30. Qiu, Y., Numerical simulation of oxygen transport in a compliant curved tube model of a coronary artery, *Ann. Biomed. Eng.*, 28, 26–38, 2000.
31. Qiu, Y. and Tarbell, J., Numerical simulation of pulsatile flow in a compliant curved tube model of a coronary artery, *J. Biomech. Eng.*, 122, 77–85, 2000.
32. Kalb, C. E. and Seader, J. D., Heat and mass transfer phenomena for viscous flow in curved circular tubes, *Int. J. Heat Mass Transfer*, 15, 801, 1972.
33. Santilli, S. M., Stevens, R. B., Anderson, J. G., Payne, W. D., and Caldwell, M. D-F., Transarterial wall oxygen gradients at the dog carotid bifurcation, *Am. J. Physiol.*, 268, H155, 1995.
34. Jo, H., Dull, R. O., Hollis, T. M., and Tarbell, J. M., Endothelial permeability is shear-dependent, time-dependent, and reversible, *Am. J. Physiol.*, 260, H1992, 1991.
35. Chang, Y. S., Yaccino, J., Lakshminarayanan, S., Frangos, J. A., and Tarbell, J. M., Shear-induced increase in hydraulic conductivity in endothelial cells is mediated is by a nitric oxide-dependent mechanism. *Arterioscler. Thromb. Vasc. Biol.*, 20, 35–42, 2000.
36. DeMaio, L., Gardner, T. G., Tarbell, J. M., and Antonetti, D. A., Shear stress-induced increase in hydraulic conductivity is correlated with increased occludin phosphorylation and decreased occulin expression in cultured endothelial monolayers. *Am. J. Physiol.*, 281H105-H113, 2001.
37. Hillsley, M. V. and Tarbell, J. M., Oscillatory shear stress alters endothelial hydraulic conductivity by a nitric oxide-dependent mechanism. *Biochem. Biophys. Res. Commun.*, 293, 1466–1471, 2002.
38. Williams, D. A., Thipakorn, B., and Huxley, V. H., *In situ* shear stress related to capillary function, *FASEB J.*, 8, M17, 1994.
39. Yuan, Y., Granger, H. J., Zawieja, D. C., and Chilian, W. M., Flow modulates coronary venular permeability by a nitric oxide-related mechanism, *Am. J. Physiol.*, 263, H641, 1992.
40. Fischer, S., Renz, D., Schaper, W., and Karliczek, G. F., Effects of barbiturates on hypoxic cultures of brain derived microvascular endothelial cells, *Brain Res.*, 707, 47, 1996.
41. Kondo, T., Kinouchi, H., Kawase, M., and Yoshimoto, T., Astroglial cells inhibit the increasing permeability of brain endothelial cell monolayer following hypoxia/reoxygeneration, *Neurosci. Lett.*, 208, 101, 1996.
42. Plateel, M., Teissier, E., and Cecchelli, R., Hypoxia dramatically increases the nonspecific transport of blood-borne proteins to the brain, *J. Neurochem.*, 68, 874, 1997.

43. Brown, L., Detmer, M., Claffey, K., Nagy, J., Peng, D., Dvorak, A., and Duorak, H., Vascular permeability factor/vascular endothelial growth factor: A multifunctional angiogenic cytokine, in: *Regulation of Angiogenesis*, Goldberg, I. D. and Rosen, E. M., Eds., Birkhäuser Verlag, Basel, Switzerland, 1997.

44. Ferrara, N. and Davis-Smyth, T., The biology of vascular endothelial growth factor, *Endocrine Rev.*, 18, 4, 1997.

45. Chang, Y., Munn, L., Hillsley, M., Dull, R., Yuan, J., Lakshminarayanan, S., Gardner, T., Jain, R., and Tarbell, J. M., Effect of vascular endothelial growth factor on cultured endothelial cell monolayer transport properties. *Microvasc. Res.*, 59, 265–277, 2000.

46. Lakshminarayaman, S., Antonetti, D., Gardner, T., and J. M. Tarbell. Effect of VEGF on hydraulic conductivity of retinal microvascular endothelial monolayers the role of NO. *Invest. Ophth. Vis. Sci.*, 41, 4256–4261, 2000.

47. Couffinhal, T., Kearney, M., Witxzenbichler, B., Chen, D., Murohara, T., Losordo, D. W., Symes, J., and Isner, J. M., Vascular endothelial growth factor/vascular permeability factor (VEGF/VPF) in normal and atherosclerotic human arteries, *Am. J. Pathol.*, 150, 1673, 1997.

48. Ramos, M. A., Kuzuya, M., Esaki, T., Miura, S., Satake, S., Asai, T., Kanda, S. Hayashi, T., and Iguchi, A., Induction of macrophage VEGF in response to oxidized LDL and VEGF accumulation in human atherosclerotic lesions, *Arterioscler. Thromb. Vasc. Biol.*, 18, 1188, 1998.

7

Transport Phenomena and the Microenvironment

7.1	Introduction 7-1
7.2	Tissue Microenvironments 7-3
	Specifying Performance Criteria • Estimating Tissue Function • Communication • Cellularity • Dynamics • Geometry • System Interactions: Reaction and Transport Processes
7.3	Reacting Systems and Bioreactors........................ 7-9
	Reactor Types • Design of Microreactors • Scale-Up and Operational Maps
7.4	Illustrative Example: Control of Hormone Diseases via Tissue Therapy 7-11
	Transport Considerations • Selection of Diabetes as Representative Case Study • Encapsulation Motif: Specifications, Design, and Evaluation
References.. 7-16

Robert J. Fisher
The SABRE Institute and Massachusetts Institute of Technology

Robert A. Peattie
Tufts University

7.1 Introduction

The evolving technologies and advances in the engineering biosciences are expected to have significant impact in the fields of pharmaceutical engineering (drug production, delivery, targeting, and metabolism), molecular engineering (biomaterial design and biomimetics), biomedical reaction engineering (microreactor design, animal surrogate systems, artificial organs, and extracorporal devices), and metabolic process control (receptor–ligand binding, signal transduction, and trafficking). Since the understanding of the cell/tissue environment will help produce major developments in all of these areas, the ability to characterize, control, and ultimately manipulate the microenvironment is critical. The key challenges, as identified by many sources (e.g., Palsson, 2000), are (1) proper reconstruction of the microenvironment for the development of tissue function, (2) scale-up to generate a significant amount of properly functioning microenvironments to be of clinical importance, (3) automating cellular therapy systems/devices to operate and perform at clinically meaningful scales, and (4) implementation in the clinical setting in concert with all the cell handling and preservation procedures required to administer cellular therapies. The direction of this chapter is toward supporting efforts to address these issues. Thus, the primary objective is to introduce the fundamental concepts needed to reconstruct tissues *ex vivo* and produce cells of sufficient quantity that maintain stabilized performance for extended time periods of clinical relevance. The delivery of cellular therapies, as a goal, was selected as one representative theme for illustration.

Before we can develop useful *ex vivo* and *in vitro* systems for the numerous applications sought, we must first have an appreciation of cellular function *in vivo*. Knowledge of the tissue microenvironment and communication with other organs is essential. We need to understand how tissue function can be built, reconstructed, and/or modified. Our approach is based on the following axioms (Palsson, 2000): (1) in organogenesis and wound healing, proper cellular communications, with respect to each other's activities, are of paramount concern since a systematic and regulated response is required from all participating cells; (2) the function of fully formed organs is strongly dependent on the coordinated function of multiple cell types with tissue function based on multicellular aggregates; (3) the functionality of an individual cell is strongly affected by its microenvironment (within 100 µm of the cell, i.e., the characteristic length scale); (4) this microenvironment is further characterized by (i) neighboring cells, that is, cell–cell contact and presence of molecular signals (soluble growth factors, signal transduction, trafficking, etc.), (ii) transport processes and physical interactions with the extracellular matrix (ECM), and (iii) the local geometry, in particular its effects on microcirculation.

The importance of the microcirculation is that it connects all the microenvironments in every tissue to their larger whole-body environment. Most metabolically active cells in the body are located within a few hundred micrometers from a capillary. This high degree of vascularity is necessary to provide the perfusion environment that connects every cell to a source and sink for respiratory gases, a source of nutrients from the small intestine, the hormones from the pancreas, liver, and glandular system, clearance of waste products via the kidneys and liver, delivery of immune system respondents, and so forth (Jain, 1994). Further, the three-dimensional (3-D) arrangement of microvessels in any tissue bed is critical for efficient functioning. This *in vivo* network develops in response to physical and chemical (molecular) clues and thus reproduction of the microenvironment with its attendant signal molecule capabilities is an essential feature of an engineered tissue system.

The engineering of these functions *ex vivo* is within the domain of bioreactor design (Freshney, 2000; Shuler, 2000), a topic discussed briefly in this chapter and elsewhere in this handbook. Cell culture devices must possess perfusion characteristics that allow for uniformity down to the 100 µm length scale. These are stringent design requirements that must be addressed with a high priority to properly account for the role of neighboring cells, the ECM, cyto-/chemokine and hormone trafficking, cell–ECM geometric factors, respiratory dynamics, and transport of nutrients and metabolic by-products for each tissue system considered. To achieve proper reconstitution of the cellular microenvironment, these dynamic, chemical, and geometric variables must be duplicated as accurately as possible. Since this is a difficult task, significant effort is devoted to developing quantitative methods to describe the cell-scale microenvironment. Once available, these methods can be used to develop an understanding of the key problems associated with any given phenomenological event, formulate solution strategies, and analyze experimental results. It is important to stress that most useful analyses in tissue engineering are performed with approximate calculations based on physiological and cell biological data, basically, determining tissue "specification sheets." Such calculations are useful for interpreting organ physiology, and providing a starting point for more extensive experimental and computational programs needed to identify the specific needs of a given tissue system (examples are given below). Using the tools obtained from studying subjects such as biomimetics (materials behavior, membrane development, and similitude/simulation techniques), transport phenomena (mass, heat, and momentum transfer), reaction kinetics, and reactor performance/design, systems that control microenvironments for *in vivo*, *ex vivo*, or *in vitro* applications can be developed.

The emphasis taken here to achieve these desired tissue microenvironments is through use of novel membrane systems designed to possess unique features for the specific application of interest, and in many cases to exhibit stimulant/response characteristics. These so-called "intelligent" or "smart" membranes are the result of biomimicry, that is, they have biomimetic features. Through functionalized membranes, typically in concerted assemblies, these systems respond to external stresses (chemical and/or physical in nature) to eliminate the threat either by altering stress characteristics or by modifying and/or protecting the cell/tissue microenvironment. An example (to be discussed further later in this chapter) is a microencapsulation motif for beta cell islet clusters to perform as an artificial pancreas. This system uses multiple membrane materials, each with its unique characteristics and performance requirements, coupled with nanospheres dispersed throughout the matrix, which contain additional materials to enhance transport and/or barrier properties, and initiate and/or respond to specific stimuli. This chapter is structured to develop an understanding of the technologies required to design systems of this nature and to ensure their stable performance.

7.2 Tissue Microenvironments

The communication of every cell with its immediate environment and other tissues establishes important spatial-temporal characteristics and develops a significant signaling/information processing network. The microenvironment is further characterized by cellular composition, the ECM, molecular dynamics (nutrients, metabolic waste products, and respiratory gases traffic in and out of the microenvironment in a highly dynamic manner), and local geometric factors (size scale of ~100 μm). Each of these can also provide the cell with important signals (dependent upon a characteristic time and length scale) to initiate specific cell functions for the tissue system to perform in a coordinated manner. If this arrangement is disrupted, cells that are unable to provide tissue function are obtained. Further discussions on this topic are presented in later sections devoted to cellular communications.

7.2.1 Specifying Performance Criteria

Each tissue or organ undergoes its own unique and complex developmental program. There are, however, a number of common features of each component of the microenvironment that will be discussed in subsequent sections. The idea is to establish general criteria to guide the design of systems possessing these requisite global characteristics and functionality. Two representative tissue microenvironments (blood and bone) are selected here for a brief comparison to illustrate common features and distinctions.

7.2.2 Estimating Tissue Function

Blood: Interpretation of the physiological respiratory function of blood has been aided by insightful yet straightforward approximating calculations to establish basic functionalities and biologic design specifications. For example, blood needs to deliver about 10 mM of oxygen per minute to the body. Given a gross circulation rate of about 5 L/min, the delivery rate to tissues is about 2 mM oxygen per L during each pass through its circulatory system. The basic requirements that circulating blood must meet to deliver adequate oxygen to tissues are determined by the following: blood leaving the lungs has a partial pressure of oxygen between 90–100 mm Hg and drops to about 35–40 mm Hg in the venous blood at rest and to about 27 mm Hg during strenuous exercise. Thus, the required oxygen delivered to the tissues is accomplished through a partial pressure drop of about 55 mm Hg, on average. Unfortunately, the solubility of oxygen in aqueous media is low; its solubility given by a Henry's law relationship, where the liquid phase concentration is linearly proportional to its partial pressure. This equilibrium coefficient is about 0.0013 mM/mm Hg. Consequently, the amount of oxygen that can be delivered by this mechanism is limited to roughly 0.07 mM, significantly below the required 2 mM. The solubility of oxygen in blood must therefore be enhanced by some other mechanism, by a factor of about 30 at rest and 60 during strenuous exercise. This, of course, is accomplished by hemoglobin within red blood cells. However, to see how this came about, let us probe a little further. Enhancement could be obtained by putting an oxygen binding protein into the perfusion fluid. To stay within the vascular bed, this protein would have to be 50–100 kDa in size. With only a single binding site, the required protein concentration is 500–1000 g/L, which is too concentrated from both an osmolarity and a viscosity (10×) standpoint and clearly impractical. Furthermore, circulating proteases will lead to a short plasma half-life for these proteins. By increasing to four sites per oxygen-carrying molecule, the protein concentration is reduced to 2.3 mM and confining it within a protective cell membrane solves the escape, viscosity, and proteolysis problems. Obviously, nature has solved these problems since these are characteristics of hemoglobin within red blood cells. Furthermore, a more elaborate kinetics study of the binding characteristics of hemoglobin shows that a positive cooperativity exists and can provide the desired oxygen transfer capabilities both at rest and under strenuous exercise.

These functions of blood establish standards that are difficult to mimic. When designing systems for *in vivo* applications promoting angiogenesis and minimizing diffusion lengths help alleviate oxygen delivery problems. Attempting to mimic this behavior in perfusion reactors, whether as extracorporeal devices or as production systems, is more complex since a blood substitute (e.g., perfluorocarbons in microemulsions) is typically needed. Performance, functionality, toxicity, and transport phenomena issues must be addressed. In summary, to maintain tissue viability and function within devices and microcapsules, methods are being developed to enhance mass transfer, especially that of oxygen. These methods include use of vascularizing membranes, *in situ* oxygen generation, use of thinner encapsulation membranes, and enhancing oxygen-carrying capacity in encapsulated materials. All these topics are addressed in subsequent sections throughout this chapter.

Bone marrow microenvironment: Perfusion rates in human bone marrow cultures are set by determining how often the media should be replenished. A dynamics similarity analysis with the *in vivo* situation is therefore appropriate. With a cell density of about 500 million cells/mL, blood perfusion through bone marrow *in vivo* is about 0.08 mL/cm^3/minute, that is, a cell-specific perfusion rate of about 2.3 mL/10 million cells/day. Cell densities on the order of one million cells/mL are typical for starting cultures; 10 million cells would be placed in 10 mL of culture media containing about 20% serum (vol/vol). To accomplish a full daily media exchange would correspond to replacing the serum at 2 mL/10 million cells/day, which is similar to the number calculated previously. These conditions were used in the late 1980s and lead to the development of prolific cell cultures of human bone marrow. Subsequent scale-up produced a clinically meaningful number of cells that are currently undergoing clinical trials.

7.2.3 Communication

Tissue development is regulated by a complex set of events in which cells of the developing organ interact with each other and with other organs and tissue microenvironments. The vascular system connects all the microenvironments in every tissue to their larger whole-body environment. As discussed previously, a high degree of vascularity is necessary to transport signal molecules through this communication network at the rate and quantity required.

Cellular communication within tissues: Cells within tissues communicate with each other for a variety of important reasons, such as localizing cells within the microenvironment, directing cellular migration, and initiating growth factor-mediated developmental programs (Long, 2000). This communication is accomplished via three primary methods: (1) cells secrete a wide variety of soluble signal and messenger molecules, including Ca^{2+}, hormones, paracrine and autocrine agents, catecholamines, growth and inhibitory factors, eicosanoids, chemokines, and many other types of cytokines, (2) via direct cell–cell contact, and (3) secreted proteins that alter the ECM chemical milieu. Each of these communication techniques differ in terms of their specificity, and their characteristic time and length scales, thus suitable to convey a particular type of message. These information exchanges are mediated by well-defined, highly specific receptor–ligand interactions that stimulate or control cell activities. For example, the appearance of specific growth factors leads to proliferation of cells expressing receptors for these growth factors. Further, chemical gradients exist that signal cells to move along tracts of molecules into a defined tissue area. High concentrations of the attractant or other signals then serve to localize cells by stopping their sojourn.

Soluble growth factors: Growth factors are a critical component of the tissue microenvironment inducing multifunctional behavior (Freshney, 2000). Their role in the signal processing network is particularly important for this chapter. They are small proteins in the size range of 15–20 kDa with a relatively high chemical stability. Initially, growth factors were discovered as active factors that originated in biological fluids and were known as colony-stimulating factors. It is now known that growth factors are produced by a signaling cell and secreted to reach target cells through autocrine and paracrine mechanisms. They bind to their receptors, found in cellular membranes, with high affinities and trigger a complex signal transduction process. Typically, the receptor complex changes in such a way that its intracellular components take on catalytic activities. These receptor–ligand complexes are internalized in some cases with a typical time constant for internalization of the order 15–30 min. It has been shown that 10,000–70,000 growth factor molecules need to be consumed to stimulate cell division in complex cell cultures. Growth factors propagate a maximum distance of about 200 μm from their secretion source. Minimum time constants for the signaling processes are about 20 min. However, much longer times are encountered if the growth factor is sequestered. The kinetics of these processes are complex and detailed analyses can be found elsewhere (Lauffenburger and Linderman, 1993) since they are beyond the scope of this chapter.

Direct cell-to-cell contact: Direct contact between adjacent cells is common in epithelially derived tissues, and can also occur with osteocytes and both smooth and cardiac myocytes. Contact is maintained through specialized membrane structures, including desmosomes, tight junctions, and gap junctions, each of which incorporates cell adhesion molecules, surface proteins, cadherins, and connexins. Tight junctions and desmosomes are thought to bind adjacent cells cohesively, preventing fluid flow between cells. *In vivo* they are found, for example, in intestinal mucosal epithelium, where their presence prevents leakage of the intestinal contents through the mucosa. In contrast, gap junctions form direct cytoplasmic bridges between adjacent cells. The functional unit of a gap junction, called a connexon, is ~1.5 nm in diameter, and thus will allow molecules below about 1 kDa to pass between cells.

These cell-to-cell connections permit mechanical forces to be transmitted through tissue beds. A rapidly growing body of literature details how fluid mechanical shear forces influence cell and tissue adhesion functions (a topic discussed more thoroughly in other chapters), and it is known that signals are transmitted to the nucleus by cell stretching and compression. Thus, the mechanical role of the cytoskeleton in affecting tissue function by transducing and responding to mechanical forces is becoming better understood.

Extracellular matrix and cell–tissue interactions: The ECM is the chemical microenvironment that interconnects all the cells in the tissue and their cytoskeletal elements. The multifunctional behavior of the ECM is an important facet of tissue performance since it provides tissue with mechanical support. The ECM also provides cells with a substrate in which to migrate, as well as serving as a storage site for signal and communications molecules. A number of adhesion and ECM receptor molecules located on the cell surface play a major role in facilitating cell–ECM communications by transmitting instructions for migration, replication, differentiation, and apoptosis. Consequently, the ECM is composed of a large number of components that have varying mechanical and regulatory capabilities that provide its structural, dynamic, and informational functions. It is constantly being modified. For instance, ECM components are degraded by metalloproteases. About 3% of the matrix in cardiac muscle is turned over daily.

The composition of the ECM determines the nature of the signals being processed and in turn can be governed or modified by the cells comprising the tissue. The ECM can direct all cellular fate processes, providing a means for cells to communicate with signals that are more stable and may be more specific and stronger than those delivered by the diffusion process needed with growth factors. A summary of the components of the ECM and their functions for various tissues is given in Palsson (2000).

Many research groups are attempting to construct artificial ECMs. The scaffolding for these matrices has taken the form of polymer materials that can be surface modified for desired functionalities. In some cases, they are designed to be biodegradable, allowing the cells to replace this material with its natural counterpart as they establish themselves and their tissue function. The major obstacle is that the properties of this matrix are difficult to specify since the properties of natural ECMs are complex and not fully known. Furthermore, two-way communication between cells is difficult to mimic since the information contained within these conversations is also not fully known. At this time, the full spectrum of ECM functionalities can only be provided by the cells themselves.

Communication with the whole-body environment: The importance of the vascular system and in particular, the microcirculation, cannot be overemphasized. This network is needed to connect all the microenvironments in every tissue to their larger whole-body environment. A complex vascular morphology is established such that most metabolically active cells in the body are located within a few hundred micrometers from a capillary. This high degree of vascularity is necessary to provide the perfusion environment that connects every cell to a source and sink for respiratory gases, a source of nutrients from the small intestine, the hormones from the pancreas, liver, and glandular system, clearance of waste products via the kidneys and liver, delivery of immune system respondents, and so forth (Freshney, 2000; Jain, 1994; other chapters in this handbook). Transport of mass (and heat) in normal and neoplastic tissues occurs primarily by convection and diffusion processes that take place throughout the whole circulatory system and ECM (Fournier, 1999; Jain, 1994). The design of *in vivo* systems therefore must consider methods to promote this communication process, not only deal with the transport issues of the devise itself. The implanted tissue system vasculature must therefore consist of (1) vessels recruited from the preexisting network of the host vasculature and (2) vessels resulting from the angiogenic response of host vessels to implanted cells (Hosack et al., 2008; Jain, 1994; Peattie et al., 2004, Riley et al., 2006). Although the implant vascular originates from the host vasculature, its organization may be completely different depending on the tissue type, its growth rate, and its location. The architecture may be different not only among varying tissue types but also between an implant and any spontaneous tissue outgrowth originating from growth factor stimuli, from the implant or as a whole-body response.

A blood-borne molecule or cell that enters the vasculature reaches the tissue microenvironment and individual cells via (i) distribution through the vascular tree, (ii) convection and diffusion across the microvascular wall, (iii) convection and diffusion through the interstitial fluid and ECM, and (iv) transport across the cell membrane (Lightfoot, 1974). The sojourn of molecules through the vasculature is governed by the morphology of that network (i.e., the number, length, diameter, and geometrical arrangement of various blood vessels) and the blood flow rate (determining perfusion performance). Transport across vessel walls to interstitial space and across cell membranes depends on the physical properties of the molecules (e.g., size, charge, and configuration), physiological properties of these barriers (e.g., transport pathways), and driving force (e.g., concentration and pressure gradients). Furthermore, specific or nonspecific binding to tissue components can alter the transport rate of molecules through a barrier by hindering the species and/or changing the transport parameters (Fisher, 1989).

Since the convective component of the transport processes via blood depends primarily on local blood flow in the tissue, coupled with vascular morphology of the tissue, hydrodynamics must be considered in designing for performance. In addition, perfusion rate requirements must take into account diffusional boundary layers along with the geometric factors of tissues and implants. In general, implant volume changes as a function of time more rapidly than for normal tissue due to tissue outgrowth, fibrotic tissue formation, and macrophage attachment. All these effects contribute to increased diffusion paths and nutrient consumption. Even with these distinctions between different tissues, the mathematical models of transport in normal, neoplastic, and implanted tissues both with and without barriers, whether *in vivo*, *ex vivo*, or *in vitro*, are identical. The only differences lie in the selection of physiological, geometric, and transport parameters. Furthermore, similar transport analyses can be applied to extracorporeal and novel bioreactor systems and their associated scale-up studies. Examples include artificial organs, animal surrogate systems, and the coupling of compartmental analysis with cell culture analogs in drug delivery and efficacy testing. Designing appropriate bioreactor systems for these applications is a challenge for tissue engineering teams collaborating with reaction engineering experts.

7.2.4 Cellularity

The number of cells found in the tissue microenvironment can be estimated as follows. The packing density of cells is on the order of a billion cells/cm^3; tissues typically form with a porosity of between 0.5 and 0.7 and therefore have a cell density of ~100–500 million cells/cm^3. Thus, an order of magnitude estimate for a cube with a 100 μm edge, the mean intercapillary length scale, is about 500 cells. For comparison, simple multicellular organisms have about 1000 cells. Of course, the cellularity of the tissue microenvironment is dependent upon the tissue and the cell types composing it. At the extreme, ligaments, tendons, aponeuroses, and their associated dense connective tissue are acellular. Fibrocartilage is at the low end of cell-containing tissues, with about one million cells/cm^3 or about 1 cell per characteristic cube. This implies that the microenvironment is simply one cell maintaining its ECM.

In most tissue microenvironments, many cell types are found in addition to the predominate cells that characterize that tissue. Leukocytes and immune system cells, including lymphocytes, monocytes, macrophages, plasma cells, and mast cells, can be demonstrated in nearly all tissues and organs, particularly during periods of inflammation. Precursor cells and residual undifferentiated cell types are present in most tissues as well, even in adults. Such cell types include mesenchymal cells (connective tissues), satellite cells (skeletal muscle), and pluripotential stem cells (hematopoietic tissues). Endothelial cells make up the wall of capillary microvessels, and thus are present in all perfused tissues.

7.2.5 Dynamics

In most tissues and organs, the microenvironment is constantly changing due to the transient nature of the multitude of events occurring. Matrix replacement, cell motion, perfusion, oxygenation, metabolism, and cell signaling all contribute to a continuous turnover. Each of these events has its own characteristic time constant. It is the relative magnitude of these time constants that dictate which processes can be considered in a pseudo-steady state with respect to the others. Determining the dynamic parameters of the major events (estimates available in Lightfoot, 1974; Long, 2000; Palsson, 2000) is imperative for successful modeling and design studies.

Time scaling of the systemic differential equations governing the physicochemical behavior of any tissue is extremely valuable in reducing the number of dependent variables needed to predict responses to selected perturbations and to evaluate system stability. In many cell-based systems, overall dynamics are controlled by transport and/or reaction rates. Controlling selected species transport then becomes the major issue since, under certain conditions, transport resistances may be beneficial. For example, when substrate inhibition kinetics is observed, performance is enhanced as the substrate transport rate is restricted. Furthermore, multiple steady states, with subsequent hysteresis problems, have been observed in these encapsulated systems as well as in continuous, suspension cell cultures (Bruns et al., 1973; Europa et al., 2000; Fisher et al., 2000). Consequently, perturbations in the macroenvironment of an encapsulated cell/tissue system can force the system to a new, less desired, steady state where cellular metabolism as measured by, for example, glucose consumption and amino acid synthesis is altered. Simply returning the macroenvironment to its original state may not be effective in returning the cellular system to its original (desired) metabolic state. The perturbation magnitudes that force a system to seek a new steady state, and subsequent hysteresis lags, are readily estimated from basic kinetics and mass transfer studies (Bruns et al., 1973; Europa et al., 2000). However, incorporation of intelligent behavior into an encapsulation system permits mediation of this behavior. This may be accomplished by controlling the cell/tissue microenvironment through the modification of externally induced chemical, biological (as in a macrophage or T cell), or physical stresses, and through selectively and temporally releasing therapeutic agents or signal compounds to modify cellular metabolism. Various novel bioreactor systems that are currently available as "off-the-shelf" items can be modified to perform these tasks in appropriate hydrodynamic flow fields, with controlled transport and/or contacting patterns, and at a microscale of relevance.

7.2.6 Geometry

Geometric similitude is important in attempting to mimic *in vivo* tissue behavior in engineered devices. The shape and size of any given tissue bed must be known to aid in the design of these devices since geometric parameters help establish constraints for both physical and behavioral criteria. Many micro-environments are effectively 2-D surfaces (Palsson, 2000). Bone marrow has a fractal cellular arrangement with an effective dimensionality of 2.7, whereas the brain is a 3-D structure. These facts dictate the type of culture technique (high density, as obtained with hollow fiber devices, versus the conventional monolayer culture) to be used for best implant performance. For example, choriocarcinoma cells release more human chorionic gonadotrophin when using this type of high-density culture (Freshney, 2000).

7.2.7 System Interactions: Reaction and Transport Processes

The interactions brought about by communications between tissue microenvironments and the whole-body system, via the vascular network, provide a basis for the systems biology approach taken to understand the performance differences observed *in vivo* versus *in vitro*. The response of one tissue system to changes in another due to signals generated, metabolic product accumulation, or hormone appearances must be properly mimicked by coupling individual cell culture analog systems through a series of micro-bioreactors if whole-body *in vivo* responses are to be meaningfully predicted. The need for microscale reactors is obvious given the limited amount of tissue or cells available for many *in vitro* studies. This is particularly true when dealing with the pancreatic system (Galletti et al., 2000; Lewis and Colton, 2004) where intact Langerhans islets must be used rather than individual beta cells for induced insulin production by glucose stimulation. Their supply is extremely limited and maintaining viability and functionality is quite complex since the islet clusters *in vivo* are highly vascularized and this feature is difficult to reproduce in preservation protocols.

The coupling of reaction kinetics with transport processes is necessary to develop effective bioreactor systems. Further discussion of this topic is given later in this chapter. See Sections 7.3 and 7.4 for reactor design specifications and mass transfer analysis in encapsulation motifs, respectively. Heat and momentum transport are major topics discussed in other chapters in this handbook. Brief comments on the necessity for these studies are presented as follows.

In all living organisms, and especially the higher animals, diffusion and flow limitations are of critical importance. In adapting problem-solving strategies for mathematical analysis and modeling of specific physiologic transport problems, it becomes particularly important to establish orders of magnitude and to make realistic limiting calculations. Especially attractive for these purposes are dimensional analysis and pharmacokinetic modeling; it seems in fact that these may permit unifying the whole of biological mass transport. Distributed-parameter transport modeling is also helpful. To obtain useful models, one must set very specific goals and work toward them by systematically comparing theory and experiment. Particularly important are the estimation of transport properties and the development of specialized conservation equations, as for ultrathin membranes, about which only limited information is known. Thus, analysis of physiologic transport stands in strong contrast to classical areas of transport phenomena (Bird et al., 2002; Brodkey and Hershey, 1988; Deen, 1996; Fournier, 1999; Lightfoot, 1974; Rosner, 1986).

Mass transfer lies at the heart of physiology and provides major constraints on the metabolic rates and anatomy of living organisms, from the organization of organ networks to molecular intracellular structures. Limitations on mass transfer rates are major constraints for nutrient supply, waste elimination, and information transmission. The primary functional units of metabolically active muscle and brain, liver lobules, and kidney nephrons have evolved to just eliminate significant mass transfer limitations in the physiological state (Lightfoot, 1974). Turnover rates of highly regulated enzymes are just less than diffusion limitations. Signal transmission rates are frequently mass transport limited (Lauffenburger and Linderman, 1993) and very ingenious mechanisms have evolved to speed these processes (Palsson, 2000). Consequently, understanding tissue mass transport is important to accurately interpret transport-based experiments that involve complex interactions of transport and reaction processes, and further, our ability to design effective devices and biosensors.

Models for microvascular heat transfer are useful for understanding tissue thermoregulation, for optimizing thermal therapies such as hypothermia treatment, for modeling thermal regulatory response at the tissue level, for assessing microenvironment hazards that involve tissue heating, for using thermal means of diagnosing vascular pathologies, and for relating blood flow to heat clearance in thermal methods of blood perfusion measurement (Jain, 1994). For example, the effect of local hypothermia treatment is determined by the length of time that the tissue is held at an elevated temperature, normally 43°C or higher. Since the tissue temperature depends on the balance between the heat added by artificial means and the tissue's ability to clear that heat, an understanding of the means by which the blood transports heat is essential for assessing the tissue's response to any thermal perturbation. The thermal regulatory system and the metabolic needs of tissues can change the blood perfusion rates in some tissues by a factor 25 (Baish, 2000). More expensive reviews and tutorials on microvascular heat transfer may be found elsewhere (Baish, 2000; Cooney, 1976; Jain, 1994).

Biological processes within living systems are significantly influenced by the flow of liquids and gases. Consequently, understanding the basic pressure and flow mechanisms involved in biofluid processes is essential for our ability to design biomimetic systems. Simulations using computational fluid dynamics (CFD) models to develop the necessary similitude and performance predictions, and the experimental evaluations of prototype devices, are crucial for successful *in vivo* implantation applications. Even for systems intended to remain *in vitro*, the effects of physiological fluids must be anticipated. Other complex devices, such as the development of bioreactors as surrogate systems, also need design guidance from CFD methods. The interaction of fluids and supported tissue is paramount in tissue engineering and the control of the microenvironment. The strength of adhesion and dynamics of detachment of mammalian cells from engineered biomaterials scaffolds is an important ongoing research, as is the effects of shear on receptor–ligand binding at cell–fluid interfaces. In addition to altering transport across the cell membrane, and possibly more important, receptor location, binding affinity and signal generation, with subsequent trafficking within the cell (Lauffenburger and Linderman, 1993), can be changed.

Furthermore, the analysis and design of reactors is highly dependent upon knowing the nonideal flow patterns that exist within the vessel. In principle, if we have a complete velocity distribution map for the fluid, then we are able to predict the behavior of any given vessel as a reactor. Once considered an impractical approach, this is now obtainable by computing the velocity distribution using CFD-based procedures in which the full conservation equations are solved for the reactor geometry of interest. Both the nonlinear nature of these equations themselves and appropriate nonlinear constitutive relationships can readily be taken into consideration.

7.3 Reacting Systems and Bioreactors

Chemical reactions, usually closely coupled with transport phenomena, sustain and support life processes. Combining these with thermodynamics identifies the area of reaction engineering. Knowledge of the fundamentals involved is essential for our understanding of these reacting systems and ability to design and control engineering devices, such as the novel flow reactor systems needed by tissue engineers. These systems provide the required experimental conditions, that is, appropriate flow fields, with controlled transport and/or contacting patterns, and at a microscale of relevance.

The coupling of encapsulation technologies with cell culture techniques permits the extended use of these bioreactors with complex tissue systems such as islet clusters. Existing reactor systems are also useful in obtaining the basic transport and reaction kinetics parameters necessary to design the novel devices required for biomedical applications. In summary, the relevance of this work to the biotechnology and health care-based industries is in the development of artificial organ systems, extracorporeal devices to bridge transplantation, biosensors, drug design, discovery, development, and controlled delivery systems. Furthermore, the control and stabilization of metabolic processes has a major impact on many research programs, including the development of animal surrogate systems for toxicity testing and biotechnological processes for the production of pharmaceuticals.

7.3.1 Reactor Types

Characterization of the mass transfer processes in bioreactors, as used in cell/tissue culture systems is essential when designing and evaluating their performance. Flow reactor systems bring together various reactants in a continuous fashion while simultaneously withdrawing products and excess reactants. These reactors generally provide optimum productivity and performance. They are classified as either tank- or tube-type reactors. Each represents extremes in the behavior of the gross fluid motion within the vessel. Tank-type reactors are characterized by instant and complete mixing of the contents and are therefore termed perfectly mixed, or back-mixed, reactors. Tube-type reactors are characterized by lack of mixing in the flow direction and are called plug flow or tubular reactors. The performance of actual reactors, though not fully represented by these idealized flow patterns, may match them so closely that they can be modeled as such with negligible error. Others can be modeled as combinations of tank and tube type over various regions.

Also in use are batch systems, where there is no flow in or out. The feed (initially charged) is placed in the reactor at the start of the process and the products are withdrawn (all at once) some time later. Reaction time is thus readily determined. This is significant since conversion is a function of time and can be obtained from the knowledge of the reaction rate and its dependence upon concentration and process variables. Since operating conditions such as temperature, pressure, and concentration (under certain circumstances) are typically controlled at a constant value, the time for reaction is the key parameter in the reactor design process. However, in a batch reactor, the concentration of reactants change with time and, therefore, the rate of reaction does as well. The performance of reactors can be predicted using rate expressions and straightforward mathematical tools.

Most actual reactors deviate from these idealized systems primarily because of nonuniform velocity profiles, channeling and bypassing of fluids, and the presence of stagnant regions caused by reactor shape and internal components such as baffles, heat transfer coils, and measurement probes. Disruptions to the flow path are common when dealing with heterogeneous systems, particularly when solids are present, as when using encapsulated systems. To model these actual reactors, various regions are compartmentalized and represented as combinations of plug flow and back-mixed elements. The study of biochemical kinetics, particularly when coupled with complex physical phenomena, such as the transport of heat, mass, and momentum, is required to determine or predict reactor performance. It is imperative to uncouple and unmask fundamental phenomenological events in reactors and to subsequently incorporate them in a concerted manner to meet objectives of specific applications. This need further emphasizes the role played by the physical aspects of reactor operation in process stability and controllability.

7.3.2 Design of Microreactors

Microscale reaction systems are often desirable for biomedical and biotechnological applications, in which the component substrates may be expensive or available only in very small quantities. This is particularly true when it is desired to conduct reactions with living cells. The advantages of microreactors are briefly summarized below. An important point to bear in mind is that these microscale systems, by their very nature, are differential reactors, that is, very low single-pass conversions. Although this single-pass analysis is quite beneficial when intermediate species kinetics studies are required, multiple passes and subsequent reservoir dynamics may be needed for higher conversion experimental programs. A few of the intrinsic qualities of microreactors are as follows: minimizes reactant (serum) requirements; minimizes the quantity of cells or biocatalysts and support material; short contact time reactions can be studied more reliably at reduced flow rate; extends the duration of experimental times, that is, longer continuous operation with small resource consumption; "aging" studies, both accelerated and long term, are more easily conducted; minimizes transport effects; coping with both mixing and flow distribution problems is less formidable and the data are more amenable to analysis; compatibility with spectroscopic systems; *in situ* studies utilizing advanced spectroscopic techniques are easier to conduct since size restrictions are reduced; and control of aseptic environment operating in a biohazard/laminar flow hood is easily implemented, as is the optimum placement of ancillary equipment.

Design criteria and performance expectations are consistent with the principles of process intensification (PI) methods. The application of this PI philosophy to numerous systems has proven its value, on both a technology and economic basis. Some insight into the virtues obtainable can be found in a description of the "bottom-up" approach to the formation of nanocrystals (Panagiotou and Fisher, 2008).

7.3.3 Scale-Up and Operational Maps

These are complex topics requiring a thorough background in chemical reaction engineering. An extensive literature has been published on the art and technology of scale-up. Here, our objective is only to direct readers to some of the sources most pertinent to bioreactors; works worthy of attention are textbooks on biomedical engineering (Fournier, 1999; Palsson, 2000) and cell culture methods (Freshney, 2000).

The use of various membrane configurations coupled with bioreactors has led to multiple functionality improvements and innovations. Implementation as guard beds, recycle conditioning vessels (with solids separations capabilities), *in situ* extraction systems, and slip-stream (and/or bypass) reactors for "biocatalyst" activity maintenance are but a few important examples representing successful applications when using living systems operating in controlled microenvironments.

The biosynthesis of lactate from pyruvate (Chen et al., 2004) is an illustrative example of the use of reactor hydrodynamic characterization and mass transport studies to improve performance through identification of operating regimes in which selected mechanisms dominate. Exploitation of this knowledge can then enhance the contribution from the desired phenomena. The operational protocols to enter these regimes, once they are documented, are termed operational maps. The multipass, dynamic input operating scheme used in that study permitted system parameter optimization studies to be conducted, including concentrations of all components in the free solution, flow rates, and electrode composition and their transport characteristics. Operating regimes that determined the controlling mechanism for process synthesis (e.g., mass transfer vs. kinetics limitations) were readily identified by varying system variables and operating conditions. Thus, procedures for developing operational maps were established and implemented.

7.4 Illustrative Example: Control of Hormone Diseases via Tissue Therapy

The treatment of autoimmune disorders with cell/tissue therapy has shown significant promise. A successful implant comprises cells or tissue surrounded by a biocompatible matrix permitting the entry of small molecules such as oxygen, nutrients and electrolytes and the exit of toxic metabolites, hormones, and other small bioactive compounds, while excluding antibodies and T cells, thus protecting the encapsulated cells/tissue. Systems of this type are currently being evaluated for the treatment of a variety of disorders, including type I diabetes mellitus, Hashimoto's disease (thyroiditis), and kidney failure. Several key issues need to be addressed before the clinical use of this technology can be realized: tissue supply, maintenance of cell viability and functionality, and protection from immune rejection. Viability and metabolic functionality are controlled by the transport of essential biochemical signal molecules, nutrients, and respiratory gases. In particular, maintenance of sufficient oxygen levels in the encapsulation device is critical to avoid local domains of necrotic and/or hypometabolic cells. Oxygen transport can be enhanced through several means, including the selection of optimal encapsulation configurations, promotion of vascularization at the implantation site, and seeding at an optimal cell density. Despite research efforts directed in these areas, oxygen transport remains one of the major limitations in maintaining cell viability and functionality. To improve this situation, recent efforts have focused on the design of a novel nanotechnology-based encapsulation motif containing specific oxygen carriers. Such a motif will ensure complete metabolic functionality of the encapsulated cells and allow the cells to retain functionality over extended time, that is, months as opposed to weeks.

7.4.1 Transport Considerations

Our emphasis here is to describe the issues relevant to development of encapsulation motifs for tissue/cellular systems that help control their microenvironments. The objective is to discuss and utilize mechanisms by which molecular transport occurs through complex media. Experimental protocols focus on the selective transport of solute molecules to evaluate proposed mechanisms and establish performance criteria. Numerous models have been postulated to explain these phenomenological observations and to develop methodologies to predict performance, thereby facilitating the design of successful encapsulation systems. Issues that need to be incorporated into these models include (1) interfacial phenomena between the bulk fluid and the outer membrane surfaces and/or along a pore wall, (2) sorption into the membrane matrix itself with diffusion possibly affected by immobilization at specific interactive sites, (3) free and/or fixed site diffusion within the matrix and if appropriate, through the porous regions, whether as distinct pores, microchannels, or other nonhomogeneous discrete areas, and (4) any chemical reactions that could alter the nature of the diffusing species or the media itself.

Two models developed previously from the analysis of transport in hydrogel membranes (Fang et al., 1998; Yasuda and Lamaze, 1971) describe the various aspects of importance to us. Both pore and sorption mechanisms may be active but their classification, and thus characterization, is based upon which, if either, dominates. When considering a pore mechanism, the solute is envisioned as passing through fluid filled micropores (or channels) in the membrane. For molecules and their mean free path much smaller than these opening, a simple Fickian diffusion model will suffice. Knudsen diffusion will be considered when the pore size is smaller than the mean free path yet still larger than molecular diameter. As the pore and molecular sizes approach each other, hindered diffusion occurs where physical/chemical interactions of significant magnitude, not simply elastic collisions, play a dominant role in transport rates. It is in this region that interactions at the molecular level, such as surface adsorption (physical or chemisorption), absorption into the matrix (solubility), and molecular transformation, become major phenomenological factors. Consideration must now be given to rate events, extent of reaction, equilibrium partitioning, irreversibility, site proximity, degree of saturation, and desorption. Migration rates along fiber or crystalline components in the matrix (and/or pore surface) are dependent upon energetics as well as site proximity, where a shunt pathway could be established to enhance transport if close enough. Otherwise, random adsorption events should hinder the transport process, analogous to diffusion with immobilization (Fisher, 1989). However, once reaction sites are fully saturated, a normal diffusive zone can be established. The influence of chemical reactive sites can be determined following prior analyses (Cussler, 1984) when appropriate, as in functionalized surfaces and/or sites distributed throughout a membrane. In contrast, in porous materials possessing microchannels, but lacking a well-defined pore structure, the dominate phase is the fluid that fills these voids. This situation is analogous to that in a swollen fibrous media, in which solute can diffuse readily but encounters fibers in its sojourn through the membrane. These encounters have a random aspect due to the nonhomogeneous nature of the membrane. They could be passive, as in elastic collisions, or active through affinity interactions as in the pore concept described above.

Transport through nonporous materials requires solute to be absorbed (solubilized) in the matrix material. Solute molecules are, thus, subject to thermodynamic equilibrium factors at the fluid–solid interfaces, as well as the nature of the fluid and solid phases themselves. These include ion strength, degree of solute hydration, and other interactive forces. Once the solute is within the membrane, a simple

Fickian diffusion mechanism may take place. Furthermore, all the other affinity events, as discussed for the pore model, could occur at interactive sites establishing a more complex transport process.

Combinations of these mechanisms may be observed in any membrane system that has distinct fluid, amorphous, crystalline, and/or functionalized regions, whether classified as porous or nonporous. Membranes may be characterized with respect to these mechanistic events, as modeled based on experimental transport measurements. The analysis tools used to interpret these results are briefly discussed later in the context of this example case study.

7.4.2 Selection of Diabetes as Representative Case Study

The National Institute of Diabetes and Digestive and Kidney Diseases (NIDDKD) estimates that there are 16 million people (nearly one in 17) that have diabetes in the United States alone. It is one of the most common and widespread diseases, with as high as 6% of the world population suffering from diabetes. In addition to the primary symptom, loss of blood glucose regulation, complications, and sequelae of diabetes include blindness, cardiovascular disease, and loss of peripheral nerve function. When including these complications, diabetes is the fourth most important cause of mortality in the United States and the main cause of permanent blindness. The American Diabetes Association estimates that diabetics consume 15% of U.S. health care costs (more than twice their percentage of the population), that is, the total cost of diabetic morbidity and mortality is more than $90 billion per year. At least 50% of that figure attributed to direct medical costs for the care of diabetic patients. Although most of the affected individuals are not dependent upon interventional insulin replacement, NIDDKD estimates that 800,000 diabetics do require this treatment to manage blood glucose regulation. Insulin delivery is not a cure, however. Restoration of normal glucose regulation by improved insulin therapy techniques that regulate insulin delivery offers the hope of circumventing the need for injection treatments and eliminating the serious debilitating secondary complications. Consequently, many research paths are being followed to determine how normal pancreatic functions can be returned to the body. These include whole-pancreas transplants, human and animal islet transplantation, fetal tissue exchange and creation of artificial beta cells, each with its pros and cons. The two major problems are the lack of sufficient organs or cells to transplant and the rejection of transplants. Since there is a severe shortage of adult pancreases, that is, 1000 patients per available organ, alternatives such as islet cell transplantation are being sought. Cell transplantation, if it could be successfully achieved, would help with both of these major problems. Use of either "artificial" islets, potentially grown from stem or beta cells themselves, or xenogenic islet clusters, in combination with designed materials for immunoisolation functionality could lead to restoration of normoperative glycemic control without the need for insulin therapy.

7.4.3 Encapsulation Motif: Specifications, Design, and Evaluation

The goal of ongoing research programs is to improve the success rate of pancreatic islet cell transplantation and thus provide a better means to regulate glucose levels for diabetic patients. This is being accomplished through immunoisolation and immunoalteration technologies implemented using intelligent membrane encapsulation systems. These systems exist as multilayered microcapsules that utilize semipermeable membranes that permit transport of nutrients, insulin, and metabolic waste products while excluding antibody and T-cell transport (Johnson et al., 2009; Lewis and Colton, 2004).

The immunoisolative capabilities of the encapsulation motif are based upon a size-exclusion principle whereby antibodies (primarily IgM and IgG) and complement proteins of the immune system are unable to reach the implanted cells. In order to activate the complement pathway in which antibodies bind specific complement proteins and ultimately destroy the implanted cells through lysis, one IgM molecule (MW = 800 kDa; ~30 nm diameter) and one molecule of complement protein C1q (MW = 410 kDa, ~30 nm diameter) must bind together. Alternatively, two IgG molecules (MW = 150 kDa each, ~20 nm total diameter) bind in concert with C1q to destroy the implant cells. Encapsulating cells in materials with a molecular weight cut-off of roughly 200 kDa therefore shields cell surface antigens from exposure to these antibodies.

It is important to also consider the ability of implanted cells to shed antigens into the surrounding host tissues, triggering another type of immune response. Many such shed antigens are comprised of major histocompatibility complex (MHC) antigens, which are too small (57–61 kDa) to be retained by the encapsulation motif. Activation of the immune system in this manner recruits macrophages, which release reactive oxygen species in an attempt to destroy the implanted cells. The inclusion of reactive sites within the matrix that function as free radical scavengers is thus desirable to protect the cells from these toxic compounds. This may be possible using nanosphere technologies, whether active ingredient-loaded or by their own characteristics, dispersing them throughout the microsized beads. Simultaneously, nanosphere technology may also be designed to augment respiratory gas exchange (Johnson et al., 2009). The solubility of O_2 in water or blood plasma is ~2 mL O_2 per 100 mL of solution at standard temperature and pressure, which is at least an order of magnitude low for encapsulation purposes. Thus, selected O_2 carriers, such as perfluorocarbons, could be incorporated into the nanospheres or as microemulsions into the matrix of the encapsulating motif.

Transport to and from encapsulated cells of nutrients, respiratory gases, and similar small molecules is not affected by pore sizes commonly found in encapsulation motifs. However, transport of desired proteins, synthesized by the cells, out of the encapsulation matrix can be limited by the pore size of the material. Large secreted proteins (e.g., MW ~ 660 kDa) will be blocked by the MW cut-offs needed to shield the implanted cells from antibodies, whereas the diffusion of MW = 28 kDa and smaller proteins will not. Consequently, the molecular weight cut-off of the encapsulation material must be carefully chosen when transport of a secreted natural compound such as insulin is desired.

The hypothesis underlying this effort is that macromolecular biomaterial encapsulation materials can be engineered that would promote islet cluster viability while simultaneously facilitating desirable biologic responses. For encapsulation materials to be physiologically functional for metabolite transport and hormone secretion, their most important properties need to be biocompatibility and selective semipermeability. However, to promote implant longevity as well as augment tissue interstitium transport and exchange characteristics, an equally important subcharacteristic is the ability to stimulate neovascularization and vascular in-growth *in situ*. Approaches such as biodegradation of a sacrificial outer layer that either releases or "generates" growth factors to promote angiogenesis are currently being studied (Hosack et al., 2008; Riley et al., 2006). Thus, a single encapsulating material will not in general possess the spectrum of properties for a successful implantation. The design of a multilayer motif, in which each layer is selected to contribute specific functions, must be considered.

One must also account for the possibility that in certain situations increased transport resistances may be beneficial. For example, when substrate inhibition kinetics behavior is observed, performance is improved as the substrate transport rate is restricted. Mammalian cells can establish multiple steady states, with subsequent hysteresis effects, while in continuous culture at the same dilution rate and feed medium. Consequently, perturbations in the macroenvironment of an encapsulated cell/tissue system can force the system to a new, less desired, steady state with altered cellular metabolism. Simply returning the macroenvironment to its original state may not be effective in returning the cellular system to its original metabolic state, and hence predicted performance is unfavorable. The magnitudes of the macroenvironment perturbations forcing the system to seek a new steady state may be estimated from models developed simulating experimentally measured metabolism. Such behavior can be mediated by the encapsulation motif through its control of the cell/tissue microenvironment. In principle, the intelligent behavior of a proposed encapsulation system will allow the maintenance of the desired microenvironment through modification of stresses and release of necessary compounds. Characterization of the required materials in this motif and subsequent efficacy testing with appropriate cell/tissue systems is an integral component of these efforts. Fortunately, the majority of materials that must meet these objectives (desired functionalities) are currently available. Examples include hydrogels and their composite systems that incorporate ionomers and other functional entities. A thorough discussion of the techniques used to characterize these materials can be found in many sources (e.g., Crank, 1956; Cussler, 1984; Deen, 1996; Sokolnicki, 2004; Sokolnicki et al., 2005). The key points are summarized as follows.

Physical and transport parameters: A variety of physical parameters may be measured or calculated as necessary for the analysis and interpretation of transport measurements. The hydrated encapsulation matrix or membrane volume is obtained using a water displacement technique. The porosity can be estimated through a simple mass balance. Membrane morphology studies are usually conducted using SEM, an available, simple, and straightforward method to determine physical characteristics of the membranes. Along with the equilibrium weight swelling ratio porosity measurement technique, surface morphology can be examined by SEM to determine the type and structure of the void space. System dimensions are obtained by direct measurement and/or simple calculations.

The membrane permeability is determined using a pseudo-steady state analysis based on Fick's law, equilibrium partitioning to the membrane surface, and the observed concentration difference across the membrane. The instantaneous flux, j, through the matrix or membrane (in a diffusion cell apparatus) is then given by

$$-j = \frac{[P]}{l}(C_D - C_R) \qquad (7.1)$$

where l is the total length of the membrane and C_D and C_R are the concentrations within the donor cell and the receptor cell, respectively (Cussler, 1984). The parameter P is the membrane permeability and is defined as $P = D \cdot H$, where the partition coefficient, H, is the ratio of solute at/on the membrane surface to those free in solution, and D is the effective diffusion coefficient in the membrane. Using a time-variant mass balance for each compartment and subject to the initial condition, $(C_{0,D} - C_{0,R}) = (C_{1,D} - C_{1,R})$ at $t = 0$, the solution obtained provides a method to interpret transport measurements and calculate the permeability:

$$P = \frac{1}{\beta t} \cdot \ln \frac{(C_{0,D} - C_{0,R})}{(C_{1,D} - C_{1,R})} \qquad (7.2)$$

The parameter β is a physical constant containing the dimensions of the diffusion cell and the membrane.

Measurements of dextran transport can be used to validate the effectiveness of a pore model for a given motif. Investigation of the relation between molecular size and rate of transport can establish whether diffusion is hindered due to pore walls or simply due to collisions with fibers in the diffusion path (Sokolnicki et al., 2005). To determine the unhindered diffusion rate of dextran molecules in water, the diffusivity at infinite dilution (Davidson and Deen, 1988; Labraun and Junter, 1994), must be calculated using the Stokes–Einstein equation:

$$D_o = kT / (6\pi\eta_w r_H) \qquad (7.3)$$

where r_H, the hydrodynamic radii of the dextran molecules, is obtained using the Stokes–Einstein correlation based on molecular weight, k is the Boltzmann's constant, T is the absolute temperature, and η_W is the viscosity of water at that temperature. Calculated diffusion coefficients of various molecular-weight dextrans in water at dilute and semidilute concentrations are reported in the literature (Callaghan and Pinder, 1983). A ratio of D_{eff} to D_o less than one indicates that a hindered diffusion process is present. If this ratio is dependent on molecular size, then pore (or microchannel) dimensions dominate versus collisions with individual fibers.

When using a horizontal diffusion cell, the diffusivity of a solute through a particular membrane can only be determined directly from these experimental transport measurements under select conditions, that is, if the partition coefficient and all physical parameters of the membrane and apparatus are known. Although membrane external dimensions can often be measured with reasonable accuracy, pore characteristics are typically lumped into an apparatus parameter that must be determined by calibration experiments and is accordingly subject to experimental error. Consequently, permeability determination is not a fundamental process. Its usefulness is restricted to applications within a data collection regime. Extensions to predict performance and provide better design protocols for novel applications requires that the fundamental parameters, diffusivity and partition coefficients, be known. Both can be obtained from desorption experiments, but data analysis from such experiments is more complicated, particularly when the motif geometry is nonspherical. One can use this technique to estimate both parameters, and then conduct adsorption tests to provide a direct measure of the partition coefficients, thereby providing redundancy checks for all three parameters, P, D, and H. Even with this more extensive data analysis program, one can only obtain effective diffusivities since the internal membrane structure is usually quite complex.

Marker species, such as vitamin B12, bovine serum albumin (BSA), and lysozyme, are generally selected to provide a reasonable range in size and properties for the solutes in desorption experiments. Membranes are initially saturated with one solute, and then immersed in a buffer solution of known volume. The mathematical analysis of the resulting desorption is based on an infinite sheet of uniform thickness placed in a solution, allowing the solute to diffuse from the sheet. Since membrane diameters are more than 100 times greater than the thickness, assumption of an infinite sheet is appropriate. The solution to this model system was developed by Crank (1956) in a form expressing the total amount of solute, M_t, in the solution at time t as a fraction of the amount after infinite time, M_∞. An infinite series form is obtained where D_{eff} appears in the exponential terms and can be recovered using a nonlinear fitting routine. The number of terms retained in the summation is dependent on the magnitude of time and the relative spacing of the system eigenvalues. The calculated diffusion coefficients are then compared to those in pure solvent and establish a basis to identify a hindered diffusion mechanism. An analysis of the adsorption behavior of the marker molecules in the membranes assists in the investigation of the mass transport phenomena by identifying if solute–matrix (fiber) interactions are significant.

The ability to execute a research program to obtain the requisite data to evaluate and implement designed encapsulation motifs, for example, develop a prototype from experimental data for clinical testing, is dependent upon coordinating all the efforts described above. This includes using the various novel bioreactor systems discussed earlier to perform these tasks in appropriate flow fields, with controlled transport and/or contacting patterns, and at a microscale of relevance. Concerted programs will help in attaining the goal of understanding the microenvironment of encapsulated systems to control and optimize tissue function.

References

Baish, J.W., 2000. Microvascular heat transfer, Chapter 8 In: *The Biomedical Engineering Handbook*, Bronzino JD (ed). 2nd Edition, CRC Press, Boca Raton, FL.

Bird, R.B., W.E. Stewart, and E.N. Lightfoot, 2002. *Transport Phenomena*, 2nd Edition, Wiley, New York.

Brodkey, R.S. and H.C. Hershey, 1988. *Transport Phenomena: A Unified Approach*. McGraw-Hill, New York.

Bruns D., J. Bailey, and D. Luss, 1973, Steady state multiplicity and stability of enzymatic reaction systems. *Biotech Bioeng*, 15, 1131.

Callaghan, P.T. and D.N. Pinder, 1983. A pulsed field gradient NMR study of self-diffusion in a polydisperse polymer system: Dextran in water. *Macromolecules*, 16, 968.

Chen, X., J.M. Fenton, R.J. Fisher and R.A. Peattie, 2004. Evaluation of in-situ electro-enzymatic regeneration of co-enzyme NADH in packed bed reactors: Biosynthesis of lactate. *J Electro Chem Soc*, 151, 2.

Cooney, D.O., 1976. *Biomedical Engineering Principles*. Dekker, New York.

Crank, J., 1956. *The Mathematics of Diffusion*. Oxford University Press, London, UK.

Cussler, E.L., 1984. *Diffusion: Mass Transfer in Fluid Systems*. Cambridge Press, New York.

Davidson, M.G. and W.M. Deen, 1988. Hindered diffusion of water-soluble macromolecules in membranes. *Macromolecules*, 21, 3474.

Deen, W.M., 1996. *Analysis of Transport Phenomena*. Oxford Press, New York.

Europa A.F., A. Grambhir, P.C. Fu, and W.S. Hu, 2000. Multiple steady states with distinct cellular metabolism in continuous culture of mammalian cells. *Biotech Bioeng*, 67, 25.

Fang, Y., Q. Cheng, and X.-B. Lu, 1998. Kinetics of *in vitro* drug release from chitosan/gelatin hybrid membranes. *J App Polymer Sci*, 68, 1751.

Fisher, R.J., 1989. Diffusion with immobilization in membranes: Transport and failure mechanisms; Part II—Transport mechanisms, In: *Biological and Synthetic Membranes*, A. Butterfield (ed). Alan R. Liss Co., New York.

Fisher R.J, S.C. Roberts, and R.A. Peattie, 2000. *Ann Biomed Eng*, 28(S1):39.

Fournier, R.L., 1999. *Basic Transport Phenomena in Biomedical Engineering*. Taylor & Francis, Philadelphia.

Freshney, R.I., 2000. *Culture of Animal Cells: A Manual of Basic Technique*, 4th Edition, Wiley-Liss, New York.

Galletti, P.M., C.K.Colton, M. Jaffrin and G. Reach, 2000. Artifical pancreas, Chapter 134 In: *The Biomedical Engineering Handbook*, 2nd Edition, J. D. Bronzino (ed). CRC Press, Boca Raton, FL.

Hosack, L.W., M.A. Firpo, J.A. Scott, G.D. Prestwich, and R.A. Peattie, 2008. Microvascular maturity elicited in tissue treated with cytokine-loaded hyaluronan-based hydrogels. *Biomaterials*, 29, 2336–47.

Jain, R.K., 1994. Transport phenomena in tumors, *Advances in Chemical Engineering*, Vol. 19, pp. 129–194, Academic Press, Inc, Orlando, FL.

Johnson, A.E., R.J. Fisher, G.C. Weir and C.K. Colton, 2009. Oxygen consumption and diffusion in assemblages of respiring spheres: Performance enhancement of a bioartificial pancreas. *Chem Eng Sci*, 64(22), 4470–87.

Lauffenburger, D.A. and J.J. Linderman, 1993. *Receptors: Models for Binding, Trafficking, and Signaling*, Oxford University Press, New York.

Lewis, A.S. and C.K. Colton, 2004. Tissue engineering for insulin replacement in diabetes, In: *Scaffolding in Tissue Engineering*, P. Ma and J. Elisseeff (eds). Marcel Dekker, New York.

Lightfoot, E.N., 1974. *Transport Phenomena and Living Systems*, John Wiley and Sons, New York.

Long, M.W., 2000. Tissue microenvironments, Chapter 118 In: *The Biomedical Engineering Handbook*, 2nd Edition, Vol. II, J. D. Bronzino (ed). CRC Press, Boca Raton, FL.

Palsson, B., 2000. Tissue engineering, Chapter 12 In: *Introduction to Biomedical Engineering*, J. Enderle, S. Blanchard and J. Bronzino (eds). Academic Press, Orlando, FL.

Panagiotou, T. and R.J. Fisher, 2008. Form nanoparticles via controlled crystallization: A bottom-up approach. *Chem Eng Prog*, 10(Oct.), 33–39.

Peattie, R.A., A.P. Nayate, M.A. Firpo, J. Shelby, R.J. Fisher, and G.D. Prestwich. 2004. Stimulation of *in vivo* angiogenesis by cytokine-loaded hyaluronic acid hydrogel implants. *Biomaterials*, 25, 2789–98.

Riley, C.M., P.W. Fuegy, M.A. Firpo, X.Z. Shu, G.D. Prestwich, and R.A. Peattie. 2006. Stimulation of *in vivo* angiogenesis using dual growth factor-loaded crosslinked glycosaminoglycan hydrogels. *Biomaterials*, 27, 5935–43.

Rosner, D.E., 1986. *Transport Processes in Chemically Reacting Flow Systems*. Butterworth Publishers, Boston.

Shuler, M.J., 2000. Animal surrogate systems, Chapter 97 In: *The Biomedical Engineering Handbook*, 2nd Edition, Vol. II, J. D. Bronzino (ed). CRC Press, Boca Raton, FL.

Sokolnicki, A.M., 2004. MS Thesis, Mass transport parameters of bacterial cellulose membranes, Tufts University, Boston, MA.

Sokolnicki, A.M., R.J. Fisher, T.P. Harrah and D.L. Kaplan, 2005. Permeability of bacterial cellulose membranes. *J Membr Sci*, 6793, 1–13.

Yasuda, H. and C.E. Lamaze, 1971. Permselectivity of solutes in homogeneous water-swollen polymer membranes. *J Macromol Sci Phys Part B*, 5, 111.

8

Transport and Drug Delivery through the Blood–Brain Barrier and Cerebrospinal Fluid

8.1   Introduction 8-1
       Barriers in the Central Nervous System • Blood–Brain
       Barrier • Transport Pathways across the Blood–Brain
       Barrier • Brain–CSF Barrier
8.2   Drug Delivery through the Blood–Brain Barrier........................ 8-7
       Permeability of the Blood–Brain Barrier • Determination of the
       Blood–Brain Barrier Permeability *In Vivo* and *Ex Vivo* • *In Vitro*
       Blood–Brain Barrier Models • Transport Models for the
       Paracellular Pathway of the Blood–Brain Barrier • Strategies
       for Drug Delivery through the Blood–Brain Barrier
8.3   Drug Delivery through the Cerebrospinal Fluid...................... 8-18
       Production and Circulation of the Cerebrospinal Fluid •
       Radioimmunotherapy Delivered through the Cerebrospinal Fluid
Acknowledgments... 8-20
References.. 8-21

Bingmei M. Fu
*The City College of the City
University of New York*

8.1 Introduction

The most complicated organ in our body is the brain. It contains 100 billion neurons and 1 trillion glial cells (nonnerve supporting cells in the brain, including astrocytes, oligodendrocytes, microglia, and ependymal cells). Along with a tremendous amount of blood vessels, these cells and surrounding extracellular matrix form highly complex, though well-organized 3-D interconnecting arrays. The movement of ions across neuronal membrane, through voltage-gated channels, conducts the information along the neuron axon at a speed up to 400 km/h. The release of neurotransmitter into the synaptic space between adjacent nerve cells mediates the communication between neurons. In order to perform its highly complicated tasks, the brain needs a substantial amount of energy to maintain electrical gradients across neuronal membranes and consequently requires a sufficient supply of oxygen and nutrients. Although it only accounts for 2% of the body weight, the brain uses 20% of the blood supply. The blood is delivered through a complex network of blood vessels that runs >650 km and passes a surface area of ~20 m². The mean distance between adjacent capillaries is ~40 µm, which allows almost instantaneous equilibration in the brain tissue surrounding the microvessels for small solutes such as glucose, amino acids, vitamins, oxygen, and so on. However, unlike peripheral microvessels in other organs where there is a relatively free small solute exchange between blood and tissue, the microvessels in the brain (cerebral microvessels) constrain the movement of molecules between blood and the brain tissue [1,2]. This unique characteristic provides a natural defense against toxins circulating in the blood, however, it also prevents the delivery of therapeutic agents to the brain.

8.1.1 Barriers in the Central Nervous System

The vascular barrier system in the brain consists of the blood–brain barrier (BBB) and the blood–cerebrospinal fluid (CSF) barriers. There is another barrier, the brain–CSF barrier, between brain tissue and the CSF. The locations of these barriers are demonstrated in Figure 8.1 [3]. The blood–brain barrier is the name for the wall of the cerebral microvessels in the brain parenchyma. At the surface of the brain parenchyma, microvessels running in the pia mater are called pial microvessels, which are often used as *in vivo* models for studying the BBB permeability. Owing to its unique structure that will be discussed in the next section, the BBB maintains very low permeability to water and solutes. In the middle of the brain parenchyma, there are ventricular cavities (ventricles) filled with CSF secreted by the epithelial cells of choroid plexus [4]. The choroid plexus is a highly vascular tissue with leaky, fenestrated capillaries covered with ependymal epithelium, which has relatively tight junctions. The multicell layer between the blood and the CSF in the choroid plexuses is called the blood–CSF barrier.

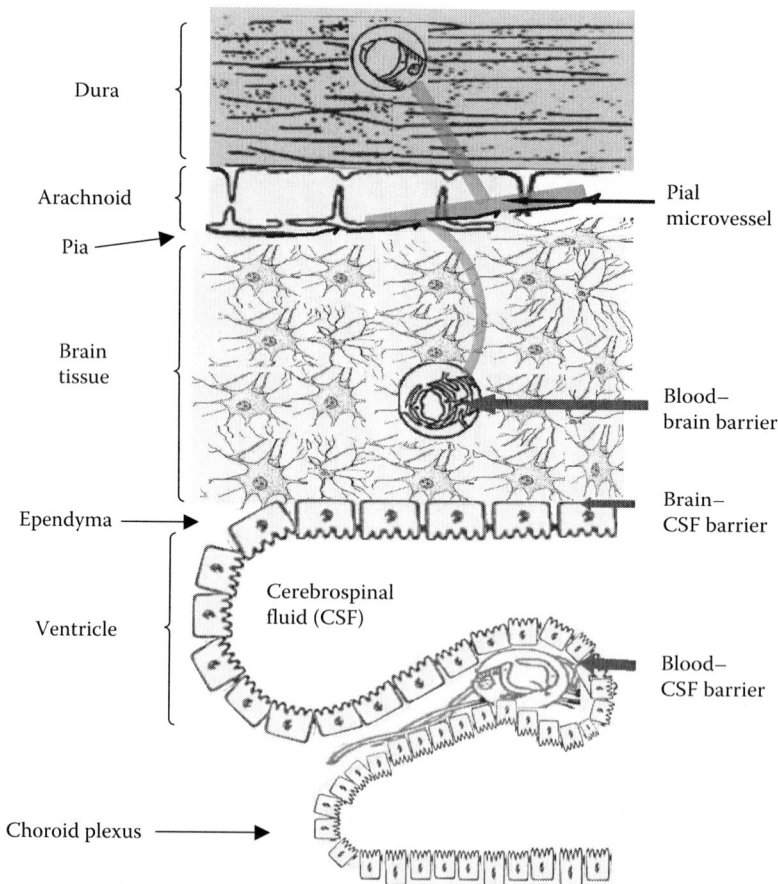

FIGURE 8.1 **(See color insert.)** Locations of barriers in the brain. (Modified from Abbott NJ, 2004. *Neurosci Int.* 45:545.)

Since the area of the BBB is about 1000 times that of the blood–CSF barrier, it is more important to circumvent the impermeability of the BBB for delivering drugs to the brain [5,6]. The total surface area of the BBB constitutes by far the largest interface for blood–brain exchange, which is between 12 and 18 m^2 for the average human adult [7]. Unlike these two tight blood barriers, the interface between the CSF and brain tissue along the ependymal surface of the ventricles and that between pia mater and brain tissue, the so-called brain–CSF barriers, are rather leaky, implying a possible route for drug delivery to the brain. The CSF is formed by the choroid plexuses of the ventricles, passes the ventricles to the subarachnoid space over the pia mater, and finally is absorbed to the venules in the dura mater through arachnoid microvilli and arachnoid granulations [8,9].

8.1.2 Blood–Brain Barrier

The BBB is a unique dynamic regulatory interface between the cerebral circulation and the brain tissue, and it is essential for maintaining the microenvironment within brain. No other body organ so absolutely depends on a constant internal microenvironment as does the brain. In the brain, the extracellular concentrations of amino acids and ions such as Na$^+$, K$^+$, and Ca^{2+} must be retained in very narrow ranges [2]. If the brain is exposed to big chemical variations in these molecules, neurons would cease to function properly because some amino acids serve as neurotransmitters and certain ions modify the threshold for neuronal firing. The BBB also protects the central nervous system (CNS) from blood-borne neuroactive solutes, such as glutamate, glycine, norepinephrine, epinephrine, and peptide hormones [10], which can increase with physiological changes (e.g., diet and stress) and pathological changes (e.g., injury and diseases). In addition, the BBB plays a key role in facilitating the brain uptake of essential nutrients like glucose, hormones, and vitamins, and larger molecules like insulin, leptin, and iron to sustain brain growth and metabolism [11].

The term "blood–brain barrier" was coined by Lewandowsky in 1900 while he demonstrated that neurotoxic agents affected brain function only when directly injected into the brain but not when injected into the systemic circulation [12]. Nevertheless, the first experimental observation of this vascular barrier between the cerebral circulation and the CNS should date back to the 1880s, when Paul Ehrlich discovered that certain water-soluble dyes, like trypan blue, after being injected into the systemic circulation, were rapidly taken by all organs except the brain and spinal cord [13]. Ehrlich interpreted these observations as a lack of the affinity of the CNS for the dyes. However, subsequent experiments performed by Edwin Goldmann, an associate of Ehrlich, demonstrated that the same dyes, when injected directly into the CNS, stained all types of cells in the brain tissue but not any other tissues in the rest of the body [14]. It took an additional 70 years until this barrier was first localized to cerebral microvascular endothelial cells by electron-microscopic studies performed by Reese and Karnovsky [15]. Although the concept of the BBB has continued to be refined over the past few decades, the recent understanding of the basic structure of the BBB is built on the general framework established by their studies in the late 1960s; more specifically, the BBB exists primarily as a selective diffusion barrier at the level of cerebral capillary endothelium.

The anatomical structure of the BBB is shown in Figure 8.2b. For comparison, the cross-sectional view of a peripheral microvessel (a typical microvessel in nonbrain organs) is also shown in Figure 8.2a. For both peripheral microvessels and the BBB, the circumference of the microvessel lumen is surrounded by endothelial cells, the opposing membranes of which are connected by tight junctions. At the luminal surface of the endothelial cell, there is a rather uniform fluffy glycocalyx layer [16–19], which is mainly composed of heparan sulfate proteoglycan, chondroitin sulfate proteoglycan, and hyaluronic acid [20]. This mucopolysaccharide structure is highly hydrated in electrolytic solution and contains large numbers of solid-bound fixed negative charges due to the polyanionic nature of its constituents abundant in glycoproteins, acidic oligosaccharides, terminal sialic acids, proteoglycan, and glycosaminoglycans aggregates [21]. Pericytes attach to the abluminal membrane of the endothelium at irregular intervals. In a peripheral microvessel, there is a loose and irregular basal lamina (or basement membrane) surrounding the pericytes. In contrast, in the BBB, pericytes and endothelial cells are ensheathed

(a)

(b)

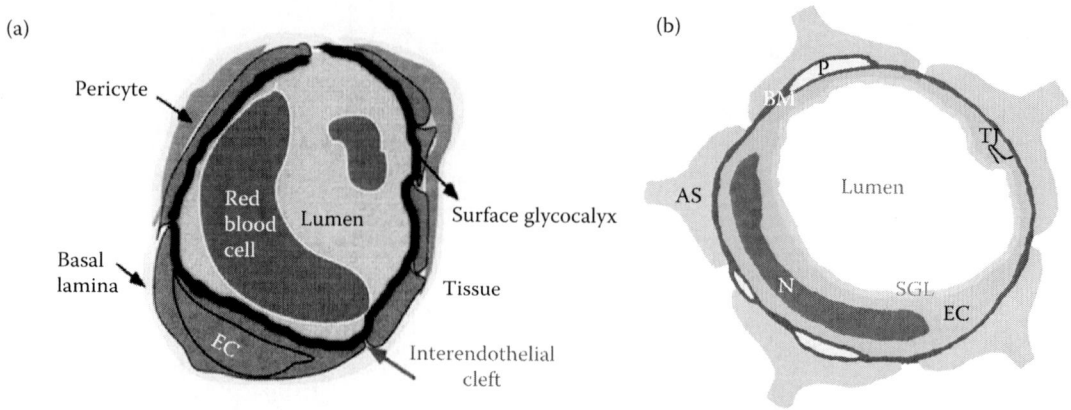

FIGURE 8.2   (**See color insert.**) Schematic of the cross-sectional view of (a) a peripheral microvessel (the microvessel in nonbrain organs), and (b) the blood–brain barrier (BBB) or cerebral microvessel (the microvessel in the brain). In addition to other structures as in a peripheral microvessel, the BBB is wrapped by astrocyte foot processes (AS, green). BM, basement membrane (or basal lamina); EC, endothelial cell; N, nucleus of endothelial cell; P, pericytes; SGL, surface glycocalyx layer; TJ, tight junction.

by a very uniform basement membrane of 20–40 nm thickness, which is composed of collagen type IV, heparin sulfate proteoglycans, laminin, fibronectin, and other extracellular matrix proteins [22]. The basal lamina is contiguous with the plasma membranes of astrocyte end-feet, which wraps almost the entire abluminal surface of the endothelium [2].

In addition to the anatomical structures, the BBB differs from the peripheral microvessels in the following aspects. The mitochondrial content of the endothelial cells forming the BBB is greater than that of such cells in all nonneural tissues. It is suggested that this larger metabolic work capacity may be used to maintain the unique structural characteristics of the BBB, and/or by metabolic pumps that may require energy to maintain the differences in composition of the cerebral circulation and the brain tissue [23]. The BBB has high electrical resistance, much less fenestration, and more intensive junctions, which are responsible for restricting paracellular passage of water and polar solutes from the peripheral circulation entering into the CNS [24,25]. Between adjacent endothelial membranes, there are junctional complexes that include adherens junctions (AJs), tight junctions (TJs), and possibly gap junctions [26]. The structure of the junction complexes between endothelial cells is shown in Figure 8.3 [27,28]. Both AJs and TJs act to restrict paracellular transport across the endothelium while gap junctions mediate intercellular communication. AJs are ubiquitous in the vasculature and their primary component is vascular endothelial (VE)-cadherin. They basically mediate the adhesion of endothelial cells to each other and contact inhibition during vascular growth and remodeling. Although disruption of AJs at the BBB can lead to increased permeability, TJ is the major junction that confers the low paracellular permeability and high electrical resistance [29]. The tight junction complex includes two classes of transmembrane molecules: occludins and claudins. These transmembrane proteins from adjacent endothelia cells interact with each other and form seals in the spaces between adjacent endothelial cells. The cytoplasmic tails of the transmembrane proteins are linked to the actin cytoskeleton via a number of accessory proteins such as members of the zonula occludens family, ZO-1, ZO-2, and ZO-3.

A number of grafting and cell culture studies have suggested that the ability of cerebral endothelial cells to form the BBB is not intrinsic to these cells, but the cellular milieu of the brain somehow induces the barrier property into the blood vessels. It is believed that all components of the BBB are essential for maintaining functionality and stability of the BBB. Pericytes seem to play a key role in angiogenesis, structural integrity, and maturation of cerebral microvessels [30]. The extracellular matrix of the basal lamina appears to serve as an anchor for the endothelial layer via interaction of laminin and other matrix proteins

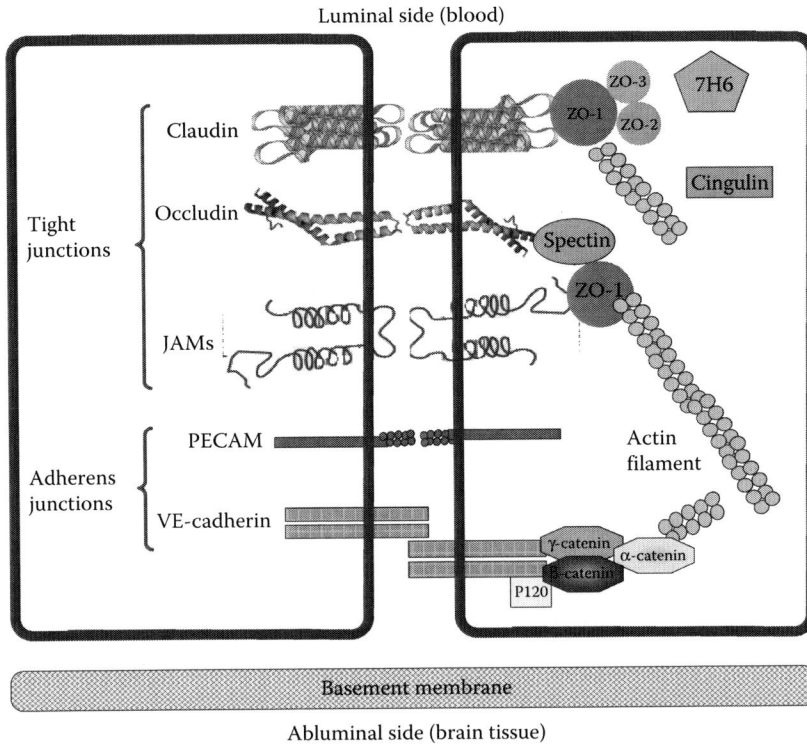

FIGURE 8.3 (**See color insert.**) Schematic of junctional complex in the paracellular pathway of the BBB. (Modified from Kim JH et al. 2006. *J Biochem Mol Biol.* 39(4):339.)

with endothelial integrin receptors [31]. It was suggested that astrocytes are critical in the development and/or maintenance of unique features of the BBB. Additionally, astrocytes may act as messengers to or in conjunction with neurons in the moment-to-moment regulation of the BBB permeability [30].

8.1.3 Transport Pathways across the Blood–Brain Barrier

The BBB endothelial cells differ from those in peripheral microvessels in that they contain more intensive tight junctions, sparse pinocytic vesicular transport, and many fewer fenestrations. The transport of substances from the capillary blood into the brain tissue depends on the molecular size, lipid solubility, binding to specific transporters, and electrical charge [32]. Figure 8.4 summarizes the transport routes across the BBB [33]. Compared to the peripheral microvessel wall, the additional structure of the BBB and tighter endothelial junctions greatly restrict hydrophilic molecules transport through the gaps between the cells, that is, the paracellular pathway of the BBB, route A in Figure 8.4. In contrast, small hydrophobic molecules such as O_2 and CO_2 diffuse freely across plasma membranes following their concentration gradients, that is, the transcellular lipophilic diffusion pathway, route C in Figure 8.4. The BBB permeability to most molecules can be estimated on the basis of their octanol/water partition coefficients [34]. For example, diphenhydramine (Benadryl), which has a high partition coefficient, can easily cross the BBB, whereas water-soluble loratadine (Claritin) is not able to penetrate the BBB and has little effect on the CNS [35].

However, the octanol/water partition coefficients do not completely reflect BBB permeability to solutes. Some solutes with low partition coefficients that easily enter into the CNS generally cross the BBB by active or facilitated transport mechanisms, which rely on ion channels, specific transporters,

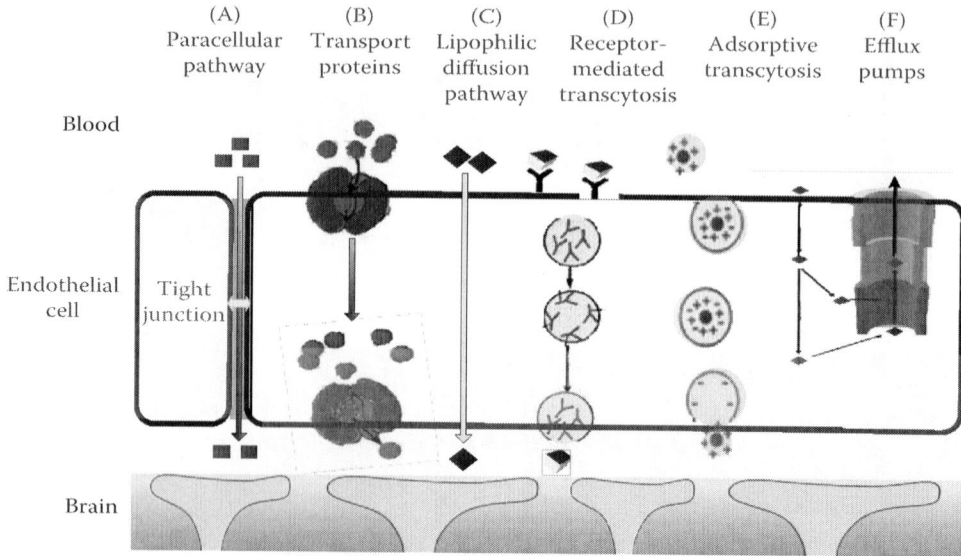

FIGURE 8.4   (See color insert.) Transport pathways across the brain endothelial cell. (Modified from Neuwelt EA, 2004. *Neurosurgery* 54(1):131.)

energy-dependent pumps, and a limited amount of receptor-mediated transcytosis (RMT). Glucose, amino acids, and small intermediate metabolites, for example, are ushered into brain tissue via facilitated transport mediated by specific transport proteins (route B in Figure 8.4), whereas larger molecules, such as insulin, transferrin, low-density lipoprotein (LDL), and other plasma proteins, are carried across the BBB via receptor-mediated (route D) or adsorptive transcytosis (route E). Some small molecules with high octanol/water partition coefficients are observed to poorly penetrate the BBB. Recent studies suggested that these molecules are actively pumped back into blood by efflux systems (route F in Figure 8.4). These efflux systems greatly limit drug delivery across the BBB. For instance, P-glycoprotein (P-gp), which is a member of the adenosine triphosphate-binding cassette family of exporters, has been demonstrated to be a potent energy-dependent transporter. P-gp contributes greatly to the efflux of xenobiotics from brain to blood and has increasingly been recognized as having a protective role and being responsible for impeding the delivery of therapeutic agents [36]. The organic anion transporters and glutathione-dependent multidrug resistance-associated proteins (MRP) also contribute to the efflux of organic anions from the CNS, and many drugs with the BBB permeability that are lower than predicted are the substrates for these efflux proteins [2,28,33,37]. While the brain endothelium is the major barrier interface, the transport activity of the surrounding pericytes [38], basement membrane, and astrocyte foot processes (Figure 8.2b) [39] also contribute to the BBB barrier function under physiological conditions, and may act as a substitute defense if the primary barrier at the endothelium is compromised [40].

8.1.4 Brain–CSF Barrier

Although the brain lacks a lymphatic drainage [41], it has a lymph-like fluid compartment, the CSF, which fills the complex interconnected ventricles within the brain, and subarachnoid space around the brain (Figure 8.5). The CSF is secreted mainly by the choroids plexus in the walls of the lateral ventricles and flows constantly and unidirectionally from the lateral ventricles through the interventricular foramina into the third, then to the fourth ventricles, finally into subarachnoid space before draining into lymphatics and veins in sagittal sinus through arachnoid villi [42]. Although the barrier between the blood and the normal brain tissue is tight and macromolecules penetration

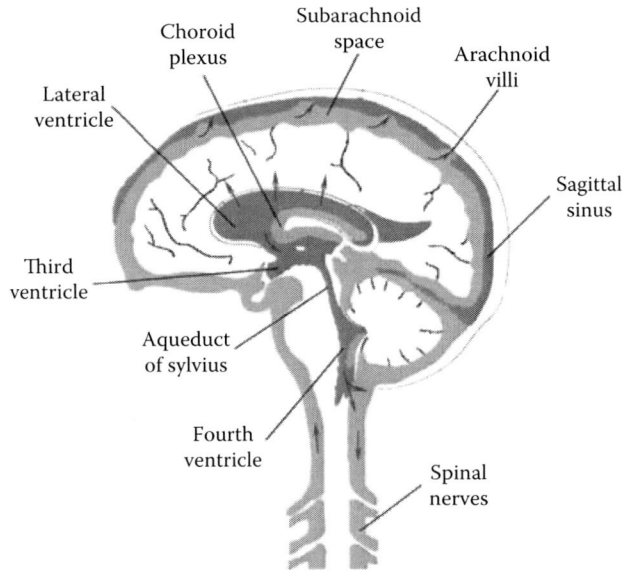

FIGURE 8.5   (**See color insert.**) Circulation of the cerebrospinal fluid (CSF) and the brain–CSF barrier. (Modified from Abbott NJ et al. 2010. *Neurobiol Dis.* 37:13.)

into CSF is hampered by tight junctions in the blood–CSF barrier [43], the transfer of macromolecules (e.g., antibodies) between the CSF in the subarachnoid space and the meninges, as well as in the ependymal lining lateral ventricles (brain–CSF barrier) is relatively free. The total exchange area of the brain–CSF barrier was estimated to be at least 1800 cm^2 [44,45]. If considering the extensive basolateral infolding (toward blood side) and the lush apical membrane microvilli (toward CSF side), the total exchange area can be the same order of magnitude as the entire BBB [46].

8.2 Drug Delivery through the Blood–Brain Barrier

The endothelial cells lining microvessel walls provide the rate-limiting barrier to extravasations of plasma components of all sizes from electrolytes to proteins. In addition to the tight junction of the microvessel endothelium, there is a uniform and narrow matrix-like basement membrane layer (20–40 nm) sandwiched between the vessel wall and the astrocyte processes ensheathing the cerebral microvessel (Figure 8.2b). To develop effective and efficient methods for drug delivery to the brain through the BBB with the largest contact area with brain tissue, we need to understand the mechanism by which these structural components, as well as transporters, receptors, efflux pumps, and other components at the endothelium and astrocyte foot processes control the permeability of the BBB to water and solutes.

8.2.1 Permeability of the Blood–Brain Barrier

Same as a peripheral microvessel, the wall of the BBB can be viewed as a membrane. The membrane transport properties are often described by Kedem–Katchalsky equations derived from the theory of irreversible thermodynamics [47]:

$$J_s = PRT\Delta C + (1 - \sigma_f)CJ_v$$
$$J_v = L_p(\Delta p - \sigma_d RT\Delta C)$$

where J_s and J_v are the solute and volumetric fluxes and ΔC and Δp are the concentration and pressure difference across the membrane. L_p, the hydraulic conductivity, describes the membrane permeability to water. P, the diffusive permeability, describes the permeability to solutes. σ_f is the solvent drag or ultrafiltration coefficient that describes the retardation of solutes due to membrane restriction, and σ_d, the reflection coefficient, describes the selectivity of membrane to solutes. In many transport processes, σ_f is equal to σ_d [47] and thus we often use σ, the reflection coefficient, to represent both of them. These three coefficients can be determined both experimentally and theoretically. In addition to these quantitative coefficients, there are other less quantitative permeability indicators for the BBB, for example, brain uptake index (BUI), and brain efflux index (BEI) [2]. In the following sections, *in vivo* and *in vitro* experiments for determining permeability of the BBB are introduced, as well as the mathematical models.

8.2.2 Determination of the Blood–Brain Barrier Permeability *In Vivo* and *Ex Vivo*

Several *in vivo* and *ex vivo* rat models have been used for the study of the transport across the BBB, including pharmacokinetic methods [48–50], intracerebral microdialysis [51], positron emission tomography (PET) [52], magnetic resonance imaging (MRI) [53], the intravital microscopy study [54], occluding single microvessel measurement [55], and single microvessel fluorescence imaging method [56].

Pharmacokinetic methods are used to evaluate the delivery of a molecule from the systemic circulation into the brain, where the amount of solute delivered to the brain can be expressed by percentage of injected dose delivered per gram of the brain. Generally, a small volume of buffered Ringer's solution containing the radiolabeled compound of interest and a radiolabeled diffusible reference compound as an internal standard (such as ^{3}H-water) is injected into the common carotid artery, or the internal carotid artery, or the venule depending on different techniques. Then, the animal is sacrificed 5–15 s after injection, and the brain tissue and the injection solution are analyzed to calculate the BUI, which is the ratio of radiolabeled test compound/^{3}H reference in the brain, divided by the ratio of radiolabeled test compound/^{3}H reference in the injection mixture [2]. Another permeability indicator, BEI, can also be determined using this method. BEI = (amount of test compound injected into the brain – amount of test compound remaining in the brain)/amount of test compound injected into the brain. The assumptions of these models are (1) the reference compound is freely diffusible across the BBB; (2) the drug does not back-diffuse from the brain to blood; and (3) no metabolism of the compounds occurs before decapitation. The advantage of these pharmacokinetic methods is speed with which many compounds can be assessed in a short time, which is ideal in the high-throughput setting. The major disadvantages are (1) brain extraction only occurs over a limited time, making it difficult to accurately determine the BUI [2] and (2) the driving force for the transport is unknown.

Intracerebral microdialysis involves direct sampling of brain interstitial fluid by a dialysis fiber implanted into the brain parenchyma. The concentration of compound that has permeated into the brain following oral, intravenous, or subcutaneous administration can be monitored over time within the same animal. Any drug that enters the brain interstitial fluid will permeate into the physiological solution within the probe, and the solution may be subsequently assayed by an appropriate technique. The major advantage of this technique is that it provides pharmacokinetic profiles of drugs in the brain without killing animals at different time points. One limitation of this technique is that it greatly relies on and is limited by the sensitivity of the assay technique [51], since the solute concentrations may be extremely low in the dialysate. Another major disadvantage is that insertion of the probe can result in chronic BBB disruption.

More recently, various imaging techniques, including PET and MRI, have been used to determine BBB permeability in humans. PET is a noninvasive tracer technique used to quantify the BBB extravasations. MRI is also a noninvasive technique, but it is more qualitative than quantitative. The major disadvantages for these techniques include their inherent costs, labor intensity, relatively low resolution (100 μm to 1 mm per pixel), and inability to differentiate between parent compound and metabolites [56,57].

All of the above-mentioned methods only measure certain indexes of relative permeability for the drug uptake to brain since they cannot determine the driving force for the efflux. Because it is hard to measure the BBB permeability in brain parenchyma, the microvessels in pia dura at the surface of brain (Figure 8.1) are often used in *in vivo* BBB permeability study. Although pial microvessels do not have the entire ensheathment of astrocytes as those cerebral microvessels in the parenchyma, the pial and cerebral microvessels appear to have many morphophysiological properties in common. These include ultrastructural characteristics, permeability of cell junctions to electron-dense tracers, transendothelium electrical resistance, and molecular properties of endothelium. For these reasons, pial microvessels are often used in the BBB permeability studies [58].

Gaber et al. [54] suggested a method to measure clearance or leakage of drug out of the pial microvessels rather than "true" permeability of the microvessels to solute. However, this method cannot determine the driving force, the concentration difference of the test solute across the BBB. The occluding single microvessel measurement is done directly on one single exposed pial microvessel after the frontal craniotomy removing a small section of the skull and the dura mater [55]. This method has well-controlled conditions, including known concentration difference, across the microvessel wall. However, recent study suggests that the exposed rat pial microvessels become leaky to both small and large molecules within 20–60 min following the craniotomy and the permeability of the exposed microvessels rises sharply after 160 min [59].

To quantify the permeability of intact rat pial microvessels and overcome the above-mentioned disadvantages, Yuan et al. [56] developed a noninvasive method, without exposing the cortex, to measure the solute permeability (*P*) of postcapillary venules on rat pia mater to various-sized solutes. The pial microvessels were observed by a high numerical aperture objective lens through a section of frontoparietal bones thinned with a microgrinder (revised surgical method from Reference 55). *P* was measured on individual pial venular microvessels with the perfused fluorescence tracer solution through the carotid artery by using highly sensitive quantitative fluorescence microscope imaging method. The major procedures are shown in Figures 8.6 and 8.7. The measured data plotted as permeability versus solute radius for the rat pial microvessels are shown in Figure 8.8. The measured permeability data for rat mesenteric microvessels are from Reference 60 for a small molecule, sodium fluorescein, an intermediate-sized molecule, α-lactalbumin, and a large molecule, albumin. Figure 8.8 shows that the solute permeability of rat pial microvessels is about an order of magnitude lower than that of rat mesenteric microvessels, from 1/11 for a small solute, sodium fluorescein, to 1/6 for a large solute, albumin or dextran 70k. The permeability data is also plotted for a small molecule, Lucifer Yellow, measured by [55,59] using an open-skull method.

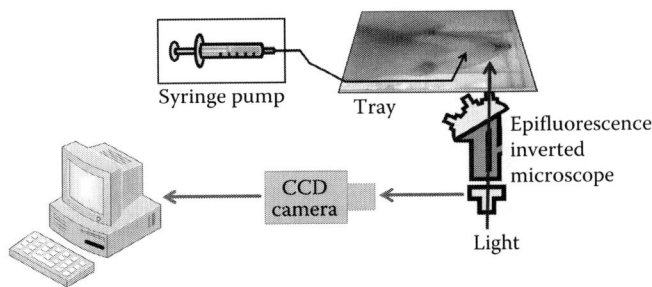

FIGURE 8.6 **(See color insert.)** Schematic for the *in vivo* permeability measurement of rat pial microvessels. The fluorescence solution was injected into the brain via a carotid artery with a syringe pump. The fluorescence images were captured by a CCD camera, which was connected to an inverted microscope. The image analysis software was then used to measure the fluorescence intensity for the region of interest in each image.

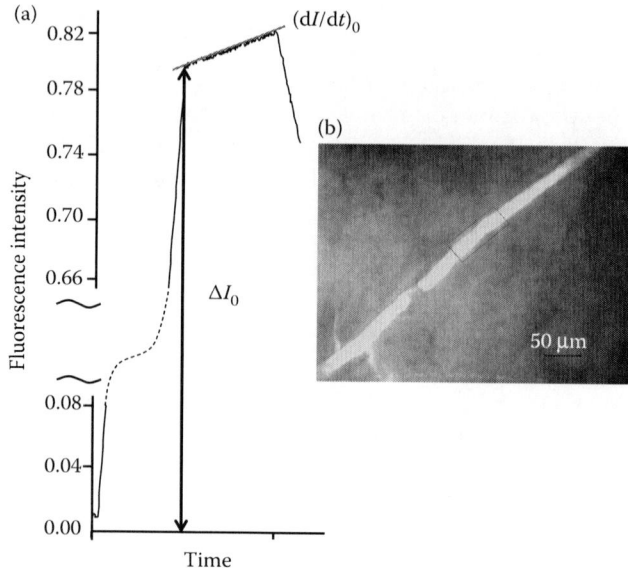

FIGURE 8.7 (**See color insert.**) Quantitative fluorescence imaging method for the measurement of solute permeability in a rat pial microvessel. The images were collected during the *in vivo* experiments and the fluorescence intensity was analyzed off-line. When the fluorescence labeled test solute was injected into the carotid artery, the pial microvessel lumen filled with fluorescent solute (red frame in b), producing ΔI_0. With continued perfusion, the measured fluorescence intensity increased indicating further transport of the solute out of the microvessel and into the surrounding tissue. The initial solute flux into the tissue was measured from the slope $(dI/dt)_0$ (a). The solute permeability P was calculated by $P = 1/\Delta I_0 \ (dI/dt)_0 \ r/2$. Here, r is the microvessel radius. The scale bar in (b) is 50 µm. (Redrawn from Yuan W et al. 2009. *Microvasc Res.* 77:166.)

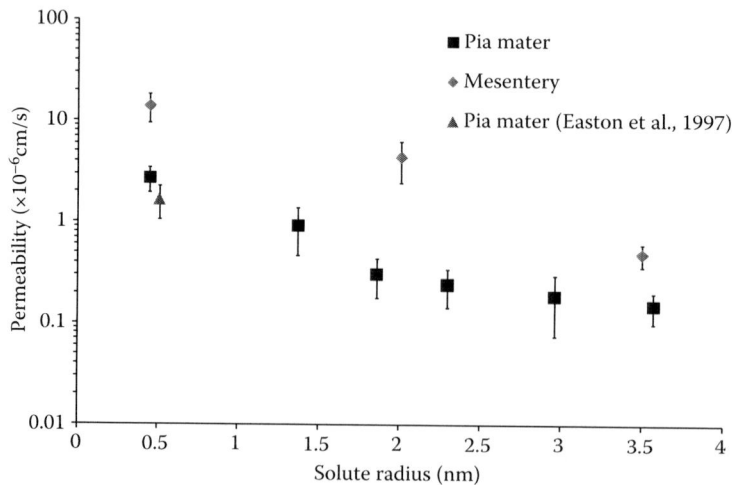

FIGURE 8.8 The pial microvessel permeability to various sized solutes versus solute radius (black squares) measured by a noninvasive method (Yuan et al., 2009) and the comparison between the measured data from Easton et al., 1997 (triangle) and those measured in mesenteric microvessels (diamonds). (Redrawn from Yuan W et al. 2009. *Microvasc Res.* 77:166.)

8.2.3 *In Vitro* Blood–Brain Barrier Models

The development of *in vitro* models for the BBB has enabled the study of transport phenomena at the molecular and cellular levels. The aim of such *in vitro* BBB models is to functionally resemble as many as possible the unique characteristics of the BBB. Compared with *in vivo* animal models, the *in vitro* models are relatively accessible, flexible, reproducible, and abundantly available. Previous investigations showed that the permeability of the *in vitro* BBB models to various compounds such as sucrose, retinoic acid, retinol, haloperidol, caffeine, and mannitol was comparable to the permeability data obtained from *in vivo* models [61].

To characterize the transport properties of *in vitro* BBB models, the solute permeability P of the *in vitro* BBB was determined by measuring the flux of the selected tracer. The most commonly used cell culture substrate consists of a porous membrane support submerged in the culture medium (Transwell apparatus). The Transwell system is characterized by a horizontal side-by-side or vertical diffusion system. During the experiment, the flux of tracers into the abluminal compartment of the Transwell system is recorded as a function of the time and the solute permeability P is calculated from the slope of the flux. The tracers used in the transport experiments are labeled by a fluorescent dye or isotope whose intensity can be measured quantitatively. Another index, transendothelial electrical resistance (TEER), or the ionic conductance of the monolayer, is also a measurement of the "tightness" of the *in vitro* BBB models.

So far, two major types of *in vitro* BBB models have been developed: endothelial cell monolayer and coculture of endothelial cells with glial cells (the nonnerve cells in the brain). The cells for these models are basically obtained from primary/subpassaged or immortalized cell cultures. The origins of the cells are also very diverse: human, primate, bovine, porcine, rodent, and murine species.

The brain capillary endothelial cells (BCEC) have been used to establish tissue culture systems ever since the technique of culturing highly purified populations of microvascular cells became available in the early 1980s. The first endothelial monolayers were established using BCEC grown on culture dishes, microcarriers (e.g., dextran beads), and various kinds of filters, including nylon mesh and polycarbonate. These cultured BCEC cells keep their endothelial phenotypes and provide a simple model for studying the permeability of the BBB. For instance, they express angiotensin-converting enzyme, von Willebrand factor, and internalize accetylated LDL. However, they were reported to lose many BBB-specific features they possessed *in vivo*. For instance, they lack specific brain endothelial markers γ-glutamyl transpeptidase, marker enzyme alkaline phosphatase [62], and glucose transporter system [63]. Moreover, the permeability of the BCEC monolayer to sucrose was reported to be from 10^{-4} to 10^{-5} cm/s compared with 10^{-6} cm/s *in vivo*. The TEER for endothelial monolayer was also found to be pretty low, from 20 to 1400 $\Omega \cdot cm^2$, compared with more than 2000 $\Omega \cdot cm^2$ *in vivo*. So the BCEC monolayer alone is not a well-characterized model for the BBB. The major reason for this may be the lack of *in situ* environment and brain-derived signals.

In the human body, the BBB is almost completely ensheathed by surrounding tissue, mostly astrocyte foot processes. Experimental results from electron microscopic techniques show that astrocytes do have significant effects on the formation of the unique BBB phenotype of brain endothelial cells [64,65]. They induce the formation of the tight junctions between endothelial cells and increase paracellular integrity of the BBB. To better mimic the *in vivo* BBB, a model with coculture of BCEC and astrocyte was developed. This coculture model was characterized on the basis of specific cell-type properties and specific BBB properties by electron microscopic evaluation and immune-histochemistry methods [66]. The results showed that BCEC displayed (1) characteristic endothelial cell morphology; (2) expression of endothelial cell markers (i.e., CD51, CD62P, CD71, and cadherin 5); (3) tight junction formation between the cells; and (4) expression of typical barrier marker γ-glutamyl-transpeptidase (γ-GTP) and P-glycoprotein (P-gp) and transferrin receptors. Astrocytes displayed characteristic astrocyte morphology and expressed glial fibrillary acidic protein (GFAP). Transmission electron microscopy showed evidence of tight junction formation between the endothelial cells and few pinocytic vesicles. A 15-fold increase in γ-glutamyl transpeptidase activity was measured in the endothelial cells cocultured with astrocytes [67]. The permeability of the coculture system to several tracers was reported to be lower than the endothelial monolayer. These results indicate that the coculture system is a better model to study the transport across the BBB.

Primary BCEC have the closest resemblance to the BBB phenotype *in vivo*, and exhibit excellent characteristics of the BBB at early passages [57]. They, however, have inherent disadvantages such as being extremely time consuming and costly to generate, being easily contaminated by other neurovascular unit cells, losing their BBB characteristics over passages, and requiring high technical skills for extraction from brain tissue [68,69]. An immortalized mouse brain endothelial cell line, bEnd3, has recently been under investigation for *in vitro* BBB models because of its numerous advantages over primary cell culture: the ability to maintain BBB characteristics over many passages, easy growth and low cost, formation of functional barriers, and amenability to numerous molecular interventions [69–73]. Previous real time polymerase chain reaction (RT-PCR) analysis showed that bEnd3 cells express the tight junction proteins ZO-1, ZO-2, occludin, and claudin-5, and junctional adhesion molecules [69,74]. They also maintained the functionality of the sodium- and insulin-dependent stereospecific facilitative transporter GLUT-1 and the P-glycoprotein efflux mechanism [74], formed fairly tight barriers to radio-labeled sucrose, and responded like primary cultures to disrupting stimuli [69].

To characterize the transport properties of *in vitro* BBB models, Malina et al. [75] and others [76–84] measured the diffusive permeability of endothelial cell monolayer and coculture of endothelial cells with astrocytes to fluorescence or isotope-labeled tracers, for example, sucrose, inulin, and mannitol. Sahagun et al. [85] reported the ratio between abluminal concentration and luminal concentration of different-sized dextrans (4k, 10k, 20k, 40k, 70k, and 150k) across mouse brain endothelial cells. Gaillard and de Boer [66] measured the permeability of sodium fluorescein and fluorescein isothiocyanate (FITC)-labeled dextran 4k across a coculture of calf BCEC with rat astrocytes. Many investigators have measured the TEER of brain endothelial monolayers and cocultures as an indicator of ion permeability [86–89].

To seek for *in vitro* BBB models that are more accessible than animals for investigating drug transport across the BBB, Li et al. [90] characterized the junction protein expression and quantified the TER and permeability to water (L_p) and solutes (P) of four *in vitro* BBB models: bEnd3 monoculture, bEnd3 coculture with astrocytes, coculture with two BM substitutes: collagen type I and IV mixture, and Matrigel. Collagen type IV network is the basic framework of native BM [91,92] and Matrigel is a soluble and sterile extract of BM derived from the EHS tumor, which has been widely used as a reconstituted BM in studying cell morphogenesis, differentiation, and growth [93]. Their results show that L_p and P of the endothelial monoculture and coculture models are not different from each other. Compared with *in vivo* permeability data from rat pial microvessels, P of the endothelial monoculture and coculture models are not significantly different from *in vivo* data for dextran 70k, but they are 2–4 times higher for small solutes TAMRA and dextran 10k. This suggests that the endothelial monoculture and all of the coculture models are fairly good models for studying the transport of relatively large solutes (drugs or drug carriers) across the BBB.

8.2.4 Transport Models for the Paracellular Pathway of the Blood–Brain Barrier

Transport across the BBB includes both paracellular and transcellular pathways [94]. While large molecules cross the BBB through transcellular pathways, water and small hydrophilic solutes cross the BBB through the paracellular pathway [25]. The paracellular pathway of the BBB is formed by the endothelial surface glycocalyx, tight junction openings, the BM filled with extracellular matrix, and the openings between adjacent astrocyte foot processes (Figure 8.2b). In addition to the endothelial tight junctions, the BM and the astrocyte foot processes provide a significant resistance to water and solute transport across the BBB.

The breakdown of the BBB and increased permeability are widely observed in many brain diseases such as stroke, traumatic head injury, brain edema, Alzheimer's disease, AIDS, brain cancer, meningitis, and so on [95–101]. Although numerous biochemical factors are found to be responsible for the breakdown of the BBB in disease, the quantitative understanding of how these factors affect the structural components of the BBB to induce BBB leakage is poor. On the other hand, the design of therapeutic drugs with better transport properties across the BBB relies greatly on this understanding. Therefore, it is important to investigate how the structural components in the paracellular pathway of the BBB affect its permeability to water and solutes through mathematical modeling.

Extended from a previous three-dimensional model for studying the transport across the peripheral microvessel wall with endothelium only [102,103], Li et al. [40] developed a new model for the transport across the BBB, which included the BM and wrapping astrocyte foot processes. The simplified model geometry is shown in Figure 8.9. This is the enlarged view for the part near the tight junction shown in Figure 8.2b. At the luminal side, there is an endothelial surface glycocalyx layer (SGL) with a thickness of L_f from 100 to 400 nm under normal physiological conditions [16–19,104]. Between adjacent endothelial cells, there is an interendothelial cleft with a length of $L \sim 500$ nm and a width of $2B \sim 20$ nm [104,105]. In the interendothelial cleft, there is a L_{jun} (~10 nm) thick junction strand with a continuous

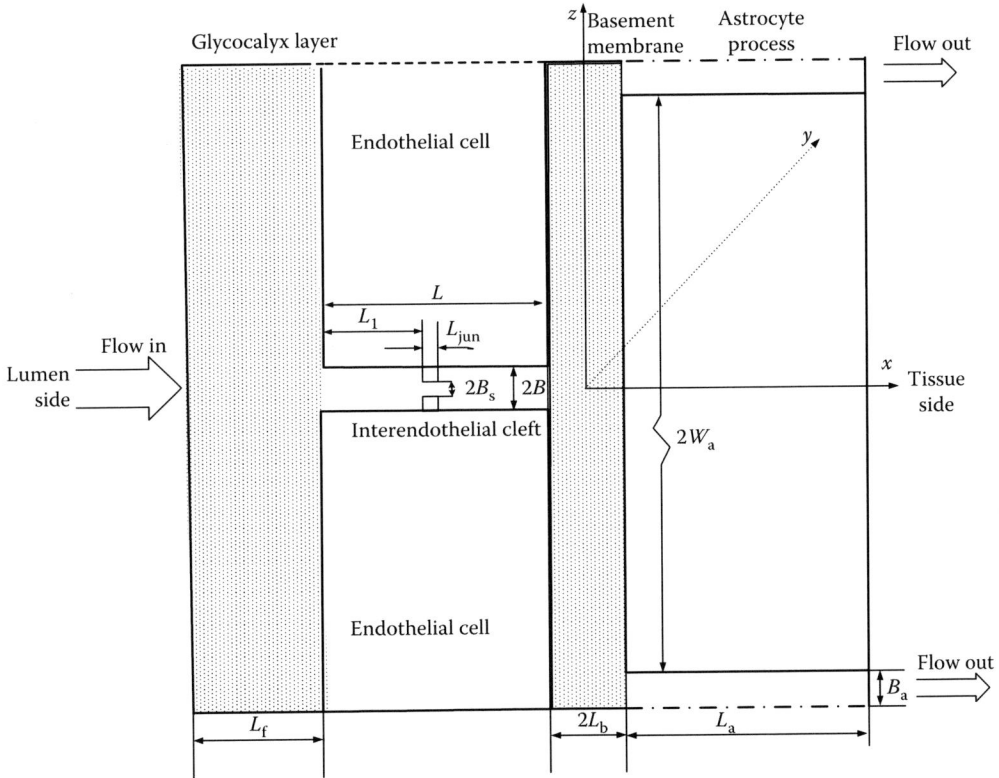

FIGURE 8.9 Model geometry for the paracellular pathway of the BBB (not in scale). The thickness of the endothelial surface glycocalyx layer is L_f. The interendothelial cleft has a length of L and a width of $2B$. The length of the tight junction strand in the interendothelial cleft is L_{jun}. The width of the small continuous slit in the junction strand is $2B_s$. The distance between the junction strand and luminal front of the cleft is L_1. The width of the basement membrane is $2L_b$ and the length of the astrocyte foot processes is $2W_a$. The cleft between astrocyte foot processes has a length of L_a and a width of $2B_a$. The surface glycocalyx layer and the endothelial cells are defined as the endothelium only while the BBB is defined to include the endothelium, the basement membrane and the astrocytes. (Redrawn from Li G, Yuan W, Fu BM, 2010. *J. Biomech.* 43:2133–2140.)

slit-like opening of width $2B_s$, which varies depending on the location of the cerebral microvessels (from ~1 to 10 nm). The distance between the junction strand and luminal front of the cleft is L_1. At the tissue side of the cleft, a BM separates the endothelium and the astrocyte foot processes. The thickness of the BM is $2L_b$ (20–40 nm) and the length of the astrocyte foot processes is $2W_a$ (~5000 nm). Between adjacent astrocyte foot processes, there is a cleft with length L_a (~1000 nm) and width $2B_a$ (20–2000 nm). The anatomic parameters for the BBB structural components were obtained from the electron microscopy studies in the literature.

Unlike the peripheral microvessel wall, the endothelium of the BBB has negligible large discontinuous breaks in the junction strand of the interendothelial cleft and the small slit in the junction strand is assumed continuous [25]. As a result, the cross-sectional BBB geometry is the same along the axial direction (y direction in Figure 8.9) and thus the model could be simplified to 2-D (in x, z plane). It could be further simplified to a unidirectional flow in each region due to very narrow clefts and the BM. In addition, the curvatures of the BM and the endothelium can be neglected because their widths are much smaller than the diameter of the microvessel. The fluid flow in the cleft regions of the BBB were approximated by the Poiseuille flow while those in the endothelial SGL and BM by the Darcy and Brinkman flows, respectively. Diffusion equations in each region were solved for the solute transport. After solving for the pressure, water velocity, and solute concentration profiles, the hydraulic conductivity L_p and solute permeability P can be calculated.

Figure 8.10a shows the model predictions for L_p as a function of tight junction opening B_s when the BM has different fiber densities. K_b is Darcy permeability in the BM. When the fiber density in the BM is the same as that in the SGL, $K_b = 3.16$ cm². The small dash-dash in Figure 8.10a shows the case of peripheral microvessels with only endothelium. When B_s increases from 0.5 to 2 nm, L_p will increase by ~20-fold. In contrast, when the endothelium is wrapped by the BM and the astrocytes as for the BBB, increase in B_s from 0.5 to 2 nm only induces fivefold increase in L_p when the fiber density in the BM is the same as that in the SGL (dash-dot-dash line). If the fiber density in the BM is 10 times of that in the SGL, the increase is only 1.6-fold in L_p (solid line), while if the fiber density in the BM is 1/10 of that in the SGL, the increase is 12-fold in L_p (large dash-dash line). Even at a large B_s of 5 nm, when the BM is filled with the same density fibers as in the SGL, the BBB permeability is only 17% of that of endothelium only. This percentage can be as low as 2% if the fiber density in the BM is 10 times of that in the SGL. Figure 8.10b shows the model predictions for L_p as a function of the endothelial SGL thickness L_f. The dashed line is for the case of endothelium only while the solid line for that of the BBB. We can see the decrease in L_f from 400 to 0 nm increases L_p by 3-fold in the case of endothelium only, while in the case of the BBB, the increase is only 25% in L_p with the protection of the BM and the astrocytes. Similar results are predicted for the solute permeability [40]. These results indicate that the BM and astrocytes of the BBB provide a great protection to the CNS under both physiological and pathological conditions. However, on the other hand, these unique structures also impede the drug delivery to the brain through the BBB.

8.2.5 Strategies for Drug Delivery through the Blood–Brain Barrier

A large number of people in the world are now suffering from CNS diseases. The total number of patients with CNS diseases is reported to be larger than that with cardiovascular diseases [5]. While the BBB serves as a natural defense that safeguards the brain against the invasion of various circulating toxins and infected cells, it also provides a significant impediment toward the delivery of diagnostic and therapeutic agents to the brain via the systemic route. Essentially, almost 100% of large-molecule drugs, including peptides, recombinant proteins, monoclonal antibodies, RNA interference (RNAi)-based drugs, and gene therapies, and more than 98% of small-molecule drugs cannot penetrate the brain microvessel wall by themselves [2,5].

Various methods such as intracerebral implantation, microdialysis, convection-enhanced distribution (CED), osmotic shock, and chemical modification of the BBB have been developed for delivering

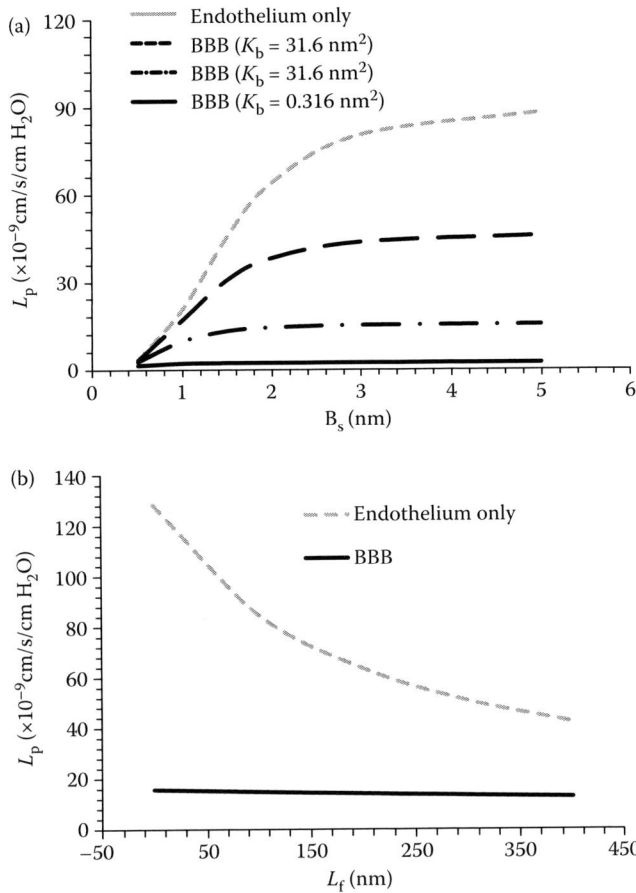

FIGURE 8.10 Model predictions for hydraulic conductivity L_p (a) as a function of B_s, the half width of the small slit in the junction strand under two cases: when considering transport across the endothelium only (endothelium only, dashed line), and when considering transport across the entire BBB (BBB). In the BBB case, three different fiber densities were considered for the basement membrane: the same as the fiber density in the surface glycocalyx layer ($K_b = 3.16$ cm², the dash-dot-dash line), 10 times lower ($K_b = 31.6$ cm², the dashed line) and higher ($K_b = 0.31$ 6cm², the solid line); (b) as a function of the surface glycocalyx layer thickness L_f. (Redrawn from Li G, Yuan W, Fu BM, 2010. *J Biomech.* 43:2133–2140.)

drugs into the brain. However, the applications of these methods are limited and they can only partially keep with the demands of modern therapies. For instance, the efficiency of intracerebral implantation, microdialysis, and CED methods are low since their major transport mechanisms are diffusion and convection of interstitial fluid. The penetration distances of drugs delivered by the first two methods are reported to be <1 mm with simple diffusion [106]. CED has been shown in laboratory experiments to deliver high-molecular-weight proteins 2 cm from the injection site in the brain parenchyma after 2 h of continuous infusion [107]. However, the success of CED relies on precise placement of the catheters and other infusion parameters for delivery into the correct location in the brain parenchyma. For effective treatment of the CNS diseases, therapeutic agents have to reach the specific regions of the brain at an adequate amount. As discussed earlier, due to the abundance and the largest contact area of the BBB for blood–brain exchange, it is more reasonable to develop strategies for drug delivery through the BBB.

As shown in Figure 8.4, we can directly deliver therapeutic agents through paracellular pathway (route A), lipophilic diffusion pathway (route C), or through transporters at the BBB by closely mimicking their substrates (route B), or mount the drugs on the ligands of the specific receptors expressed at the BBB (e.g., LDL receptor-related protein) for transcytosis (Trojan horse approach or RMT route D), as well as using cationized proteins, peptides, and nanoparticle carriers for adsorptive-mediated transcytosis (AMT, route E). The following summarizes the delivery strategies through these routes, respectively.

8.2.5.1 Delivery through Paracellular Pathway (Route A)

To increase the hydrophilic drug delivery from the blood to the brain tissue, we can transiently open the barriers in the paracellular pathway of the BBB, for example, the cleft opening ($2B$ in Figure 8.9), the tight junction opening ($2B_s$), the BM width ($2L_b$), or degrade the fiber matrix in the endothelial surface glycocalyx and in the BM. Osmotic shock by intracarotid administration of a hyperosmotic mannitol causes endothelial cells to shrink and increase $2B$, $2B_s$, and $2L_b$. Subsequent administration of drugs can increase their concentrations in the brain to a therapeutic level [108,109]. Physical means such as application of electric and magnetic fields can increase the drug brain uptake. Focused ultrasound, guided by MRI, combined with microbubbles injected into the blood stream has been shown to disrupt the BBB and increase the distribution of Herceptin in brain tissue by 50% in mice [110–112]. Application of inflammatory agents such as bradykinin analog can open the tight junction of the BBB and increase the drug transport to the brain [113,114]. However, these approaches are relatively costly and nonpatient friendly. They may also enhance tumor dissemination after BBB disruption and damage the neurons by allowing the passage of unwanted blood components into the brain [115].

8.2.5.2 Delivery through Lipophilic Diffusion Pathway (Route C)

Some molecules, for example, alcohol, nicotine, and benzodiazepine, can freely enter the brain through route C in Figure 8.4. Their ability to passively (diffusion by concentration differences across the cell membrane) cross the BBB depends on the molecular weight (<500 Da), charge, (low hydrogen bonding capabilities) and lipophilicity [116]. Therefore, if we can modify the drugs through medicinal chemistry, for example, reduce the relative number of polar groups, or incorporate them with a lipid carrier, we can enhance their brain uptake [2,117]. Modification of antioxidants with pyrrolopyrimidines increases their ability to access target cells in the CNS [118]. Covalently attaching 1-methyl-1,4-dihydronicotinate to a hydroxymethyl group can enhance the delivery of ganciclovir (Cytovene, an antiviral medication) to the brain [119,120]. However, the modification that helps for the drug delivery to the brain often results in loss of the therapeutic function of a drug. In addition, increase of lipophilicity of a drug can result in making it a substrate for the efflux pump P-glycoprotein (route F in Figure 8.4) [115].

8.2.5.3 Delivery through Transporter-Mediated Pathway (Route B)

The brain requires a tremendous amount of essential substances for survival and function, for example, glucose, insulin, hormones, and LDL. These nutrients and substances are transported into the brain, not by paracellular or lipophilic diffusion pathway as described earlier, but by specific transporters or receptors at the BBB. Drugs can be modified to take advantage of the native BBB nutrient transporter systems, or by being conjugated to ligands that recognize receptors expressed at the BBB for the RMT. This physiological approach is by far recognized as the most likely successful drug delivery method to the brain.

Peptides and small molecules may use specific transporters expressed on the luminal and basolateral sides of the endothelial cells to cross into the brain. So far, at least eight different nutrient transporters have been identified to transport a group of nutrients with similar structures. Drugs can be modified to closely mimic the endogenous carrier substrates of these transporters and be transported through the specific transporter-mediated transcytosis. Dopamine can be used to treat Parkinson's disease, but itself is nonbrain penetrant. Instead, dopamine's metabolic precursor, L-Dopa, if delivered by a neutral amino acid carrier through its transporter at the BBB, shows a clear clinical benefit on patients with Parkinson's disease [6,121]. To use a BBB transporter for drug delivery, several important factors must be considered: the kinetics and structural binding requirements of the transporter, therapeutic compound manipulation so that the compound binds but also remains active *in vivo*, and actual transport of the compound into the brain instead of just binding to the transporter [115].

8.2.5.4 Delivery through Receptor-Mediated Pathway (Route D)

Instead of by transporters, larger essential molecules are delivered into the brain by specific receptors highly expressed at the endothelial cells of the BBB. The RMT includes three steps: receptor-mediated endocytosis of the molecule at the luminal side of the endothelium (blood side), transport through the endothelial cytoplasm, and exocytosis of the molecule at the abluminal side of the endothelium (brain side). Although the exact mechanisms of RMT have not been well understood, drug delivery targeting three receptors, the insulin receptor, the transferrin receptor, and the LDL receptor, has been in use since the start of this century [121,122]. More and more receptors have been targeted for the drug delivery since then [115]. This physiological approach is often called a molecular Trojan horse since the therapeutic compounds are conjugated to the specific ligands or the antibodies, which can be recognized and delivered through transcytosis by the specific matching receptors at the endothelial cell membrane. In addition to molecular Trojan horses, drugs can be packaged to liposomes and other nanoparticles coated with targeting molecules such as antibodies to the specific receptors to improve the drug-loading capacity.

Although the Trojan horses for the BBB drug delivery are very promising in delivering large peptides and recombinant proteins such as neurotrophins, enzymes, and monoclonal antibodies [122], the traffic is limited by the number and carrying capacity of the receptors, as well as by the number of drug molecules that can attach to each antibody [123]. In addition, Gosk et al. [124] showed that using anti-transferrin mAb for drug delivery through the systemic administration, although the total amount of the drug in the brain is high, most of it stays associated with brain microvessel endothelial cells instead of in the brain parenchyma. Due to the high affinity of the antibodies, it is a challenge to dissociate from their specific receptors. Furthermore, widespread distribution of the receptors on peripheral organs would limit the specific delivery to the brain, and on the other hand, may induce additional toxicity.

8.2.5.5 Delivery through Adsorptive-Mediated Pathway (Route E)

AMT involves endocytosis and exocytosis of charged substances by the endothelial cells of the BBB. Its mechanism is different from that of the RMT, which needs specific matching receptors and ligands. Kumagai et al. [125] observed that polycationic proteins such as protamine could not only bind to the endothelial cell surface but also penetrate the BBB. Mixing protamine, poly-l-lysine or other cationic molecules with proteins (e.g., albumin) largely increased the BBB permeability to these proteins. These findings can be explained by AMT triggered by electrostatic interactions between the positively charged proteins and negatively charged membrane regions at the brain endothelium. At normal physiological pH, the luminal surface of the cerebral endothelium and the surrounding BM (see Figure 8.2b) carry negative charge [126] and provide the necessary environment for delivering positively charged drugs and drug carriers. Recently, a quantitative *in vivo* animal study by Yuan et al. [127] found that the charge density of the endothelial surface glycocalyx and that of the BM in rat pial microvessels are ~30 mEq/L. In another *in vitro* cell culture study, Yuan et al. [72] found the similar charge density on the surface of a cell monolayer of bEnd3, an immortalized mouse cerebral microvessel endothelial cell line.

To efficiently deliver a therapeutic protein or peptide across the BBB, the simplest way is to cationize the protein or peptide by amidation of its carboxylic acid groups, as well as glutamic and aspartic acid side chain groups with positively charged amines [128]. The degree of cationization of a protein or peptide may be critical for its pharmacokinetic fate. Cationization enhances delivery while inducing potential toxicity and immunogenicity of these proteins. PEGylation of cationized molecules can minimize the immunogenicity of these molecules. Positively charged cell-penetrating peptides (CPPs) are often used as the drug carriers for the brain delivery. Commonly used CPPs are penetratin, transportan, Syn-B, and Tat [128]. Brain uptake of enkephalin analog dalargin was enhanced several hundred fold after they were carried by the CPPs [129]. Decoration of CPPs on the surface of liposome and biopolymer-based nanoparticles containing drugs has been shown to promote their uptake by the brain and entrance to the cytoplasm of neurons [130]. The drawbacks through AMT are lack of tissue selectivity, although the BBB may contain higher concentrations of negative charges than other tissues, and possible disruption of the BBB and binding of polycationic substances to the negatively charged plasma proteins and other anionic sites resulting in toxicity [131].

8.3 Drug Delivery through the Cerebrospinal Fluid

As described in the introduction, unlike the BBBs that have very low permeability to water and solutes, the brain–CSF barrier (see Figure 8.1), the interface between the brain tissue and CSF, is relatively leaky, allowing passage of macromolecules (e.g., antibodies) at certain degree. Although the exchange area of the brain–CSF barrier was estimated to be at least 1800 cm^2, the total exchange area can be the same order of magnitude as the entire BBB if the infolding and microvilli of the ependyma (the epithelial cells forming the brain–CSF barrier) are taken into consideration. Therefore, delivery through the brain–CSF barrier is very promising for treating certain types of brain diseases.

8.3.1 Production and Circulation of the Cerebrospinal Fluid

The CSF is actively secreted by ependymal cells, the epithelial cells lining the highly vascularized choroid plexus (see Figures 8.1 and 8.5). The rate of secretion in humans is ~500 mL/day. The CSF circulates from the lateral ventricles through the interventricular foramina into the third ventricle, then down the narrow aqueduct of sylvius into the fourth ventricle, where it exits through two lateral apertures and one median aperture. It then flows through the cerebellomedullary cistern down the spinal cord and into the subarachnoid space over the cerebral hemispheres (Figure 8.5). Finally, the CSF passes from the subarachnoid space into venous blood of sinus sagittalis superior in dura mater through arachnoid villi and arachnoid granulations. Drainage into the lymphatic system also occurs [8,42].

The human brain contains only 135–150 mL of CSF, so the CSF flow is rapid due to a high secretion rate of 500 mL/day. The high producing rate is achieved because the structure of choroid plexus is superbly adapted to the secretion [9]. First, the choroid plexus has abundant blood supply. Compared to the average blood supply in the cerebral cortex, which is 0.35–0.4 mL/min/g in a rat, the blood supply to the lateral ventricle choroid plexus is about ~4 mL/min/g, a 10-fold more [132]. Second, in contrast to the cerebral capillaries (the BBB), the capillaries in the choroid plexus are fenestrated and provide much less resistance to the substances transport from the blood to the epithelial cells lining the choroid plexus. Third, the choroid epithelium is a leaky type of epithelium and permits the material passage between the interstitial space and the CSF. Fourth, the choroidal epithelial cells possess numerous microvilli on the apical (CSF) side and extensive infolding at the basolateral (blood) side, which increase the surface area for contact between epithelia and the interstitium at one side, and epithelia and the CSF at the other side. Fifth, the choroid epithelial cells have rich mitochondria, suggesting that the production of ATP is sufficient to sustain normal secretion of CSF. Because of its rapid refreshing rate, broad contact area with the brain tissue, and relatively leaky barrier between the CSF and the brain, the CSF is becoming a very good candidate for drug delivery to the CNS system.

8.3.2 Radioimmunotherapy Delivered through the Cerebrospinal Fluid

For delivery through the CSF, drugs can be infused intraventricularly using a plastic reservoir called an Ommaya reservoir implanted subcutaneously in the scalp and connected to the ventricles within the brain via an outlet catheter [133] (Figure 8.11). Compared to vascular drug administration, CSF delivery

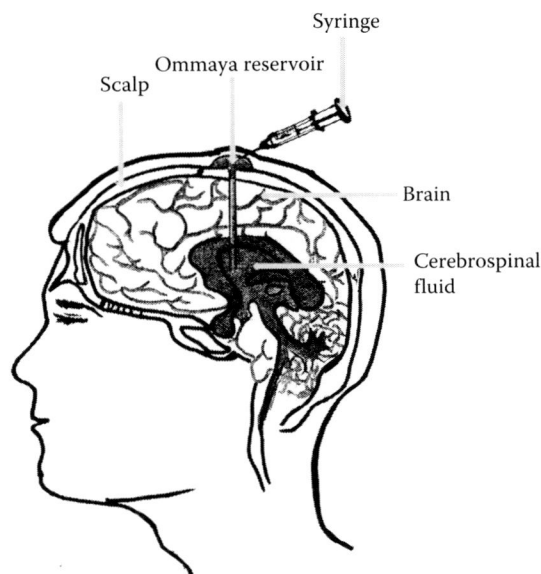

FIGURE 8.11 Drug delivery to the brain through cerebrospinal fluid.

has several advantages: first, it bypasses the BBB and immediately enhances the CSF drug concentrations; second, it potentially reduces the systemic toxicity since the drug directly goes to the CNS; third, the half life of a drug can be longer since there is less enzymatic degrading activity in the CSF; fourth, the CSF can freely exchange molecules with the extracellular fluid of the brain parenchyma, delivering drugs into the CSF is more likely resulting in therapeutic CNS drug concentrations. The following presents a successful example for brain cancer treatment by delivering isotope-labeled therapeutic antibodies (radioimmunotherapy) through the CSF.

Tumor cells can invade the CSF and disseminate throughout the neuroaxis by the constant flow of CSF which travels from the ventricles to the spinal canal and over the cortical convexities. Involvement of the leptomeninges (LM) by any cancer is a serious complication with significant morbidity and mortality [134–136]. Its frequency is increasing as patients live longer and as neuro-imaging modalities improve, approaching 5% in solid tumors such as breast cancer and lung cancer [135]. Neuroblastoma is the most common extracranial tumor of the sympathetic nervous system, occurring predominantly in early childhood and accounting for 6.7% of childhood cancer. With increasing period of remission, CNS metastasis (both parenchymal and LM), though rare in the past [136], has substantially increased in the past decade. Antibody-based radioimmunotherapy (RIT) administered through the CSF has clinical potential in the treatment of cancers metastatic to the LM or brain. ^{131}I-labeled monoclonal antibodies (mAbs) targeting GD2 (e.g., mAb 3F8) or B7H3 (e.g., mAb 8H9) when administered through an Ommaya reservoir have proven safe in phase I clinical trials [137]. Patients with relapsed neuroblastoma in the CNS (brain or LM), when treated with salvage regimens containing either intra-Ommaya ^{131}I-3F8 or ^{131}I-8H9, have survived for extended periods of time.

Given the unique physiology of the CSF compartment, as well as the well-defined kinetic/radiochemical properties of monoclonal antibodies, RIT delivered through the CSF can be optimized. An optimization model will help understand the complex dynamics of antibody and radiation dose delivered to tumor cells and to normal brain, and provide a tool to define the critical parameters in order to improve effectiveness and safety of ^{131}I-MoAb RIT administered through the CSF. In addition, since different isotopes have distinct microdosimetric properties, their biologic effect in CSF RIT can be simulated and compared. These critical parameters of CSF dynamics and mAb pharmacokinetics can

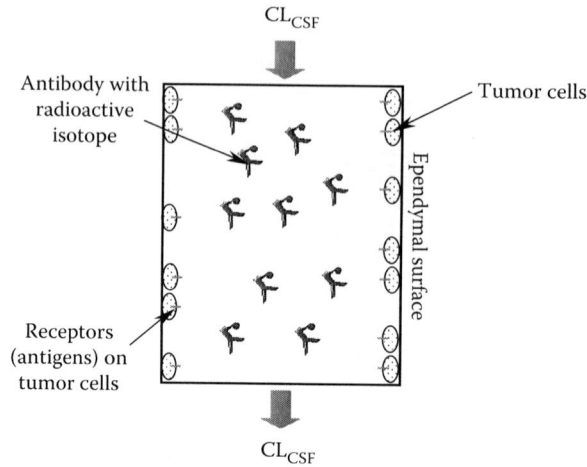

FIGURE 8.12 Schematic of pharmacokinetic model for radioimmunotherapy delivered through CSF administration. CL_{CSF} represents the flow rate of the cerebrospinal fluid. (Redrawn from Lv Y, Cheung NK, Fu BM, 2009. *J Nuclear Med.* 50(8):1324.)

be manipulated by pharmacologic interventions and genetic engineering, respectively, and be tested *in vivo* using rat or mice with LM xenografts. The long-term plan is to bring these concepts to human phase I and II studies.

Recently, Lv et al. [138] developed a pharmacokinetic model to evaluate the role of kinetic and transport parameters of RIT in maximizing the therapeutic ratio, the ratio of the area under the curve for the concentration of the bound antibodies versus time [$AUC(C_{IAR})$], to that for unbound antibodies [$AUC(C_{IA})$]. The CSF space was simplified as a single compartment and considered the binding of antibodies to antigens on tumor cells lining the surface of the CSF space (Figure 8.12). Mass conservation was applied to set up the equations for C_{IAR}, C_{IA}, and other pharmacokinetic variables and a Runge–Kutta method was used to solve the equations. This model agreed with the measured data in 10 of 14 patients in the phase I trial of intra-Ommaya RIT using ^{131}I-3F8 [133]. Using this model, several predictions were made: (1) Increasing the affinity of antibodies to antigens greatly increases $AUC(C_{IAR})$ but not $AUC(C_{IA})$. (2) For the same amount of isotope administered, the smaller antibody dosage and the higher specific activity improves therapeutic ratio. (3) When the isotope half life is 0.77 h, increasing the antibody association constant enhances $AUC(C_{IAR})$ much more than decreasing the dissociation constant even if overall affinity is unchanged. However, when the half life reaches 240 h, decreasing dissociation constant would slightly enhance $AUC(C_{IAR})$. (4) Decreasing the CSF bulk flow rate increases $AUC(C_{IAR})$. (5) At the same amount of antibody administered by continuous infusion and by split administrations, one can improve $AUC(C_{IAR})$ by up to 1.8-fold and 1.7-fold, respectively, compared to that by the single bolus administration. The improved therapeutic ratio by optimized kinetic and transport parameters may enhance clinical efficacy of this new treatment modality through the CSF.

Acknowledgments

The author would like to thank the funding support from NSF CBET-0754158, NIH 1SC1GM086286-01A1, and 1U54CA137788-01. She would also like to thank Dr. Guanglei Li, Dr. Wei Yuan, Ms Lingyan Shi, and Ms Lin Zhang for their assistance in preparing Figures 8.1, 8.2b, 8.3 through 8.6, and 8.11.

References

1. Abbott NJ, 1992. Comparative physiology of the blood-brain barrier. In: Bradbury MWB (ed) *Physiology and Pharmacology of the Blood-Brain Barrier.* Springer, Heidelberg. p. 371.
2. Pardridge WM, 1998. CNS drug design based on principles of blood-brain barrier transport. *J Neurochem.* 70:1781.
3. Abbott NJ, 2004. Evidence for bulk flow of brain interstitial fluid: Significance for physiology and pathology. *Neurosci Int.* 45:545.
4. Brown PD, Davies SL, Speake T, Millar ID, 2004. Molecular mechanisms of cerebrospinal fluid production. *Neurosci.* 129:957.
5. Pardridge WM, 2006. *Introduction to the Blood-Brain Barrier: Methodology, Biology and Pathology.* Cambridge University Press, Cambridge.
6. Pardridge WM, 2007. Drug targeting to the brain. *Pharm Res.* 24:1733.
7. Nag S, Begley DJ, 2005. Blood-brain barrier, exchange of metabolites and gases. In: Kalimo H (ed) *Pathology and Genetics. Cerebrovascular Diseases.* ISN Neuropath. Press, Basel. p. 22.
8. Culter RWP, Page L, Galicich J, Watters GV, 1968. Formation and absorption of cerebrospinal fluid in man. *Brain* 91:707.
9. Redzic ZB, Segal MB, 2004. The structure of the choroid plexius and the physiology of the choriod plexius epithelium. *Adv Drug Del Rev.* 56:1695.
10. Smith QR, 2000. Transport of glutamate and other amino acids at the blood-brain barrier. *J Nutr.* 130:1016.
11. Zhang Y, Pardridge WM, 2001. Rapid transferrin efflux from brain to blood across the blood-brain barrier. *J Neurochem.* 76:1597.
12. Lewandowsky M, 1900. Zur lehre von der cerebrospinalflussigkeit. *Z Klin Med.* 40:480.
13. Ehrlich P, 1885. *Das Sauerstufbudurfnis Des Organismus.* Hireschwald. Berlin.
14. Goldmann E, 1913. Vitalfarbung am zentralnervensystem. *Abhandl Kongil Preuss Akad Wiss.* 1:1.
15. Reese TS, Karnovsky MJ, 1967. Fine structural localization of a blood-brain barrier to exogenous peroxidase. *J Cell Biol.* 34(1):207.
16. Adamson RH, Clough G, 1992. Plasma proteins modify the endothelial cell glycocalyx of frog mesenteric microvessels. *J Physiol.* 445:473.
17. Squire JM, Chew M, Nneji G, Neal C, Barry J, Michel C, 2001. Quasi-periodic substructure in the microvessel endothelial glycocalyx: A possible explanation for molecular filtering? *J Struct Biol.* 136(3):239.
18. Ueno M, Sakamoto H, Liao YJ, Onodera M, Huang CL, Miyanaka H, Nakagawa T, 2004. Blood-brain barrier disruption in the hypothalamus of young adult spontaneously hypertensive rats. *Histochem Cell Biol.* 122(2):131.
19. Vink H, Duling BR, 1996. Identification of distinct luminal domains for macromolecules, erythrocytes, and leukocytes within mammalian capillaries. *Circ Res.* 79(3):581.
20. Henry CB, Duling BR, 1999. Permeation of the luminal capillary glycocalyx is determined by hyaluronan. *Am J Physiol Heart Circ Physiol* 277(2):H508.
21. Tarbell JM, Pahakis MY, 2006. Mechanotransduction and the glycocalyx. *J Int Med.* 259(4):339.
22. Farkas E, Luiten PGM, 2001. Cerebral microvascular pathology in aging and Alzheimer's disease. *Prog Neurobiol.* 64(6):575.
23. Oldendorf WH, Cornford ME, Brown WJ, 1977. The large apparent work capability of the blood-brain barrier: A study of the mitochondrial content of capillary endothelial cells in brain and other tissues of the rat. *Ann Neurol.* 1:409.
24. Butt AM, Jones HC, Abbott NJ, 1990. Electrical resistance across the blood-brain barrier in anaesthetized rats: A developmental study. *J Physiol.* 429:47.
25. Hawkins BT, Davis TP, 2005. The blood-brain barrier/neurovascular unit in health and disease. *Pharmacol Rev.* 57:173.

26. Simard M, Arcuino G, Takano T, Liu QS, Nedergaard M, 2003. Signaling at the gliovascular interface. *J Neurosci.* 23(27):9254.

27. Kim JH, Park JA, Lee SW, Kim WJ, Yu YS, Kim KW, 2006. Blood-neural barrier: Intercellular communication at glio-vascular interface. *J Biochem Mol Biol.* 39(4):339.

28. Abbott NJ, Patabendige AAK, Dolman DEM, Yusof SR, Begley DJ, 2010. Structure and function of the blood-brain barrier. *Neurobiol Dis.* 37:13.

29. Romero IA, Radewicz K, Jubin E, Michel CC, Greenwood J, Couraud PO, Adamson P, 2003. Changes in cytoskeletal and tight junctional proteins correlate with decreased permeability induced by dexamethasone in cultured rat brain endothelial cells. *Neurosci Lett.* 344(2):112.

30. Ballabh P, Braun A, Nedergaard M, 2004. The blood-brain barrier: An overview: Structure, regulation, and clinical implications. *Neurobiol Dis.* 16:1.

31. Hynes RO, 1992. Integrins: Versatility, modulation, and signaling in cell adhesion. *Cell.* 69:11.

32. Moody DM, 2006. The blood-brain barrier and blood-cerebral spinal fluid barrier. *Semin Cardiothorac Vasc Anesth.* 10(2):128.

33. Neuwelt EA, 2004. Mechanisms of disease: The blood-brain barrier. *Neurosurgery* 54(1):131.

34. Sawchuk RJ, Elmquist WF (2000) Microdialysis in the study of drug transporters in the CNS. *Adv Drug Deliv Rev.* 45:295.

35. Kay GG, 2000. The effects of antihistamines on cognition and performance. *J Allergy Clin Immunol.* 105:S622.

36. Schuetz EG, Schinkel AH, Relling MV, Schuetz JD, 1996. P-glycoprotein: A major determinant of rifampicin-inducible expression of cytochrome P4503A in mice and humans. *Proc Natl Acad Sci USA.* 93:4001.

37. Begley DJ, 2007. Structure and function of the blood-brain barrier. In: Touitou E, Barry BW (eds) *Enhancement in Drug Delivery.* CRC Press. Boca Raton, FL. p. 575.

38. Shimizu S, 2008. A novel approach to the diagnosis and management of meralgia paresthetica. *Neurosurgery.* 63(4):E820.

39. Wolburg-Buchholz K, Mack AF, Steiner E, Pfeiffer F, Engelhardt B, Wolburg H, 2009. Loss of astrocyte polarity marks blood-brain barrier impairment during experimental autoimmune encephalomyelitis. *Acta Neuropathol.* 18(2):219.

40. Li G, Yuan W, Fu BM, 2010. A model for water and solute transport across the blood-brain barrier. *J. Biomech.* 43:2133–2140.

41. Davson H, Segal MB, 1995. *Physiology of the CSF and Blood-Brain Barriers.* CRC Press, Boca Raton, FL.

42. Guyton AC, Hall JE, 2000. *Textbook of Medical Physiology.* 10th Edition. Philadelphia: Saunders. p. 709.

43. Smith DE, Johanson CE, Keep RF, 2004. Peptide and peptide analog transport systems at the blood-CSF barrier. *Adv Drug Deliv Rev.* 56:1765.

44. Barta P, Dazzan P, 2003. Hemispheric surface area: Sex, laterality and age effects. *Cereb Cortex.* 13:364.

45. Blasberg RG, Patlak CS, Shapiro WR, 1977. Distribution of methotrexate in the cerebrospinal fluid and brain after intraventricular administration. *Cancer Treat Rep.* 61:633.

46. Keep RF, Jones HC, 1990. A morphometric study on the development of the lateral ventricle choroid plexus, choroid plexus capillaries and ventricular ependyma in the rat. *Brain Res Dev Brain Res.* 56:47.

47. Curry FE, 1984. Mechanics and thermodynamics of transcapillary exchange. In: Renkin EM and Michel CC (eds.) *Handbook of Physiology. The Cardiovascular System.* The American Physiology Society, Bethesda. Sect.2 (vol IV, pt1):309.

48. Cornford EM, Young D, Paxton JW, Sofia RD, 1992. Blood-brain barrier penetration of felbamate. *Epilepsia.* 33:944.

49. van Uitert RL, Sage JI, Levy DE, Duffy TE, 1981. Comparison of radio-labeled butanol and iodoantipyrine as cerebral blood flow markers. *Brain Res.* 222:365.

50. Zlokovic BV, Begley DJ, Djuricic BM, Mitrovic DM, 1986. Measurement of solute transport across the blood-brain barrier in the perfused guinea pig brain: Method and application to *N*-methyl-alpha-aminoisobutyric acid. *J Neurochem.* 46:1444.

51. de Lange EC, de Boer BA, Breimer DD, 1999. Microdialysis for pharmacokinetic analysis of drug transport to the brain. *Adv Drug Deliv Rev.* 36:211.

52. Elsinga PH, Hendrikse NH, Bart J, Vaalburg W, van Waarde A, 2004. PET Studies on P-glycoprotein function in the blood-brain barrier: How it affects uptake and binding of drugs within the CNS. *Curr Pharm Des.* 10:1493.

53. Wang R, Ashwal S, Tone B, Tian HR, Badaut J, Rasmussen A, Obenaus A, 2007. Albumin reduces blood-brain barrier permeability but does not alter infarct size in a rat model of neonatal stroke. *Pediatr Res.* 62:261.

54. Gaber MW, Yuan H, Killmar JT, Naimark MD, Kiani MF, Merchant TE, 2004. An intravital microscopy study of radiation-induced changes in permeability and leukocyte-endothelial cell interactions in the microvessels of the rat pia mater and cremaster muscle. *Brain Res Brain Res Protoc.* 13:1.

55. Easton AS, Fraser PA, 1994. Variable restriction of albumin diffusion across inflamed cerebral microvessels of the anaesthetized rat. *J Physiol.* 475:147.

56. Yuan W, Lv Y, Zeng M, Fu BM, 2009. Non-invasive measurement of solute permeability in cerebral microvessels of the rat. *Microvasc Res.* 77:166.

57. Nicolazzo JA, Charman SA, Charman WN, 2006. Methods to assess drug permeability across the blood-brain barrier. *J Pharm Pharmacol.* 58:281.

58. Allt G, Lawrenson JG, 1997. Is the pial microvessel a good model for blood-brain barrier studies? *Brain Res Rev.* 24:67.

59. Easton AS, Sarker MH, Fraser PA, 1997. Two components of blood-brain barrier disruption in the rat. *J Physiol.* 503 (Pt 3):613.

60. Fu BM, Shen S, 2004. Acute VEGF effect on solute permeability of mammalian microvessels in vivo. *Microvasc Res.* 68:51.

61. Franke H, Galla HJ, Beuckmann CT, 1999. An improved low-permeability in vitro-model of the blood-brain barrier: Transport studies on retinoids, sucrose, haloperidol, caffeine and mannitol. *Brain Res.* 818(1):65.

62. Meyer J, Mischeck U, Veyhl M, Henzel K, Galla HJ, 1990. Blood-brain barrier characteristic enzymatic properties in cultured brain capillary endothelial cells. *Brain Res.* 514(2):305.

63. Hemmila JM, Drewes LR, 1993. Glucose transporter (GLUT1) expression by canine brain microvessel endothelial cells in culture: An immunocytochemical study. *Adv Exp Med Biol.* 331:13.

64. Abbott NJ, 2002. Astrocyte-endothelial interactions and blood-brain barrier permeability. *J Anat* 200(6):629.

65. Haseloff RF, Blasig IE, Bauer HC, Bauer H, 2005. In search of the astrocytic factor(s) modulating blood-brain barrier functions in brain capillary endothelial cells in vitro. *Cell Mol Neurobiol.* 25(1):25.

66. Gaillard PJ, de Boer AG, 2000. Relationship between permeability status of the blood-brain barrier and *in vitro* permeability coefficient of a drug. *Eur J Pharm Sci.* 12(2):95.

67. Demeuse P, Kerkhofs A, Struys-Ponsar C, Knoops B, Remacle C, de Aguilar PV, 2002. Compartmentalized coculture of rat brain endothelial cells and astrocytes: A syngenic model to study the blood-brain barrier. *J Neurosci Methods.* 121(1):21.

68. Deli MA, Abraham CS, Kataoka Y, Niwa M, 2005. Permeability studies on *in vitro* blood-brain barrier models: Physiology, pathology, and pharmacology. *Cell Mol Neurobiol.* 25(1):59.

69. Brown RC, Morris AP, O'Neil RG, 2007. Tight junction protein expression and barrier properties of immortalized mouse brain microvessel endothelial cells. *Brain Res.* 1130(1):17.

70. Soga N, Connolly JO, Chellaiah M, Kawamura J, Hruska KA, 2001. Rac regulates vascular endothelial growth factor stimulated motility. *Cell Commun Adhes.* 8(1):1.

71. Yoder EJ, 2002. Modifications in astrocyte morphology and calcium signaling induced by a brain capillary endothelial cell line. *Glia* 38(2):137.

72. Yuan W, Li G, Fu BM, 2010. Effect of surface charge of immortalized mouse cerebral endothelial cell monolayer on transport of charged solutes. *Ann Biomed Eng.* 38(4):1463.

73. Tyagi N, Moshal KS, Sen U, Vacek TP, Kumar M, Hughes WM Jr., Kundu S, Tyagi SC, 2009. H2S protects against methionine-induced oxidative stress in brain endothelial cells. *Antioxid Redox Signal* 11(1):25.

74. Omidi Y, Campbell L, Barar J, Connell D, Akhtar S, Gumbleton M, 2003. Evaluation of the immortalised mouse brain capillary endothelial cell line, bEnd3, as an *in vitro* blood-brain barrier model for drug uptake and transport studies. *Brain Res.* 990(1–2):95.

75. Malina KC, Cooper I, Teichberg VI, 2009. Closing the gap between the in-vivo and in-vitro blood-brain barrier tightness. *Brain Res.* 1284:12.

76. Bowman P D, Ennis SR, Rarey KE, Betz AL, Goldstein GW, 1983. Brain microvessel endothelial cells in tissue culture: A model for study of blood-brain barrier permeability. *Ann Neurol.* 14(4):396.

77. Thompson SE, Cavitt J, Audus KL, 1994. Leucine-enkephalin effects on paracellular and transcellular permeation pathways across brain microvessel endothelial-cell monolayers. *J Cardiovasc Pharmacol.* 24(5):818.

78. Salvetti F, Cecchetti P, Janigro D, Lucacchini A, Benzi L, Martini C, 2002. Insulin permeability across an *in vitro* dynamic model of endothelium. *Pharm Res.* 19(4):445.

79. Karyekar CS, Fasano A, Raje S, Lu RL, Dowling TC, Eddington ND, 2003. Zonula occludens toxin increases the permeability of molecular weight markers and chemotherapeutic agents across the bovine brain microvessel endothelial cells. *J Pharm Sci.* 92(2):414.

80. Hamm S, Dehouck B, Kraus J, Wolburg-Buchholz K, Wolburg H, Risau W, Cecchelli R, Engelhardt B, Dehouck MP, 2004. Astrocyte mediated modulation of blood-brain barrier permeability does not correlate with a loss of tight junction proteins from the cellular contacts. *Cell Tissue Res.* 315(2):157.

81. Kemper EM, Boogerd W, Thuis I, Beijnen JH, van Tellingen O, 2004. Modulation of the blood-brain barrier in oncology: Therapeutic opportunities for the treatment of brain tumours? *Cancer Treatment Rev.* 30(5):415.

82. Boveri M, Berezowski V, Price A, Slupek S, Lenfant AM, Benaud C, Hartung T, Cecchelli R, Prieto P, Dehouck MP, 2005. Induction of blood-brain barrier properties in cultured brain capillary endothelial cells: Comparison between primary glial cells and C6 cell line. *Glia* 51(3):187.

83. Kraus J, Voigt K, Schuller AM, Scholz M, Kim KS, Schilling M, Schabitz WR, Oschmann P, Engelhardt B, 2008. Interferon-beta stabilizes barrier characteristics of the blood-brain barrier in four different species in vitro. *Multiple Sclerosis* 14(6):843.

84. Poller B, Gutmann H, Krahenbuhl S, Weksler B, Romero I, Couraud PO, Tuffin G, Drewe J, Huwyler J, 2008. The human brain endothelial cell line hCMEC/D3 as a human blood-brain barrier model for drug transport studies. *J Neurochem.* 107(5):1358.

85. Sahagun G, Moore SA, Hart MN, 1990. Permeability of neutral vs. anionic dextrans in cultured brain microvascular endothelium. *Am J Physiol.* 259(1 Pt 2):H162.

86. Santaguida S, Janigro D, Hossain M, Oby E, Rapp E, Cucullo L, 2006. Side by side comparison between dynamic versus static models of blood-brain barrier in vitro: A permeability study. *Brain Res.* 1109:1.

87. de Vries HE, BlomRoosemalen MCM, van Oosten M, de Boer AG, van Berkel TJC, Breimer D D, Kuiper J, 1996. The influence of cytokines on the integrity of the blood-brain barrier in vitro. *J Neuroimmunol.* 64(1):37.

88. Cucullo L, McAllister MS, Kight K, Krizanac-Bengez L, Marroni M, Mayberg MR, Stanness KA, Janigro D, 2002. A new dynamic *in vitro* model for the multidimensional study of astrocyte-endothelial cell interactions at the blood-brain barrier. *Brain Res.* 951(2):243.

89. Zhang Y, Li CSW, Ye YY, Johnson K, Poe J, Johnson S, Bobrowski W, Garrido R, Madhu C, 2006. Porcine brain microvessel endothelial cells as an *in vitro* model to predict *in vivo* blood-brain barrier permeability. *Drug Metab Disp.* 34(11):1935.

90. Li G, Simon M, Shi Z, Cancel L, Tarbell JM, Morrison B, Fu BM, 2010. Permeability of endothelial and astrocyte cocultures: *In vitro* blood-brain barrier models for drug delivery. *Ann Biomed Eng.* 38(8):2499.

91. Engvall E, 1995. Structure and function of basement membranes. *Int J Dev Biol* 39(5):781.

92. Miosge N, 2001. The ultrastructural composition of basement membranes in vivo. *Histol Histopathol.* 16(4):1239.

93. Kleinman HK, Martin GR, 2005. Matrigel: Basement membrane matrix with biological activity. *Semin Cancer Biol.* 15(5):378.

94. Pardridge WM, 2005. Molecular biology of the blood-brain barrier. *Mol Biotechnol.* 30(1):57.

95. Dietrich WD, Alonso O, Halley M, 1994. Early microvascular and neuronal consequences of traumatic brain injury: A light and electron microscopic study in rats. *J Neurotrauma.* 11(3):289.

96. Fukuda K, Tanno H, Okimura Y, Nakamura M, Yamaura A, 1995. The blood-brain barrier disruption to circulating proteins in the early period after fluid percussion brain injury in rats. *J Neurotrauma.* 12(3):315.

97. Baldwin SA, Fugaccia I, Brown DR, Brown LV, Scheff SW, 1996. Blood-brain barrier breach following cortical contusion in the rat. *J Neurosurg.* 85(3):476.

98. Barzo P, Marmarou A, Fatouros P, Corwin F, Dunbar J, 1996. Magnetic resonance imaging-monitored acute blood-brain barrier changes in experimental traumatic brain injury. *J Neurosurg.* 85(6):1113.

99. Baskaya MK, Rao AM, Dogan A, Donaldson D, Dempsey RJ, 1997. The biphasic opening of the blood-brain barrier in the cortex and hippocampus after traumatic brain injury in rats. *Neurosci Lett.* 226(1):33.

100. Beaumont A, Marmarou A, Hayasaki K, Barzo P, Fatouros P, Corwin F, Marmarou C, Dunbar J, 2000. The permissive nature of blood brain barrier (BBB) opening in edema formation following traumatic brain injury. *Acta Neurochir Suppl.* 76:125.

101. Cernak I, Vink R, Zapple DN, Cruz MI, Ahmed F, Chang T, Fricke ST, Faden AI, 2004. The pathobiology of moderate diffuse traumatic brain injury as identified using a new experimental model of injury in rats. *Neurobiol Dis.* 17(1):29.

102. Fu BM, Chen B, 2003. A model for the structural mechanisms in the regulation of microvessel permeability by junction strands. *ASME J Biomech Eng.* 125:620.

103. Fu BM, Tsay R, Curry FE, Weinbaum S, 1994. A junction-orifice-entrance layer model for capillary permeability: Application to frog mesenteric capillaries. *ASME J Biomech Eng.* 116:502.

104. Schulze C, Firth JA, 1992. Interendothelial junctions during blood-brain-barrier development in the rat—Morphological-changes at the level of individual tight unctional contacts. *Dev Brain Res* 69(1):85.

105. Adamson RH, Lenz JE, Zhang X, Adamson GN, Weinbaum S, Curry FE, 2004. Oncotic pressures opposing filtration across non-fenestrated rat microvessels. *J Physiol-Lond.* 557:889.

106. Mak M, Fung L, Strasser JF, Saltzman WM, 1995. Distribution of drugs following controlled delivery to the brain interstitium. *J Neurooncol* 26(2):91.

107. Bobo RH, Laske DW, Akbasak A, Morrison PF, Dedrick RL, Oldfield EH, 1994. Convection-enhanced delivery of macromolecules in the brain. *Proc Natl Acad Sci USA.* 91(6):2076.

108. Kroll RA, Neuwelt EA, 1998. Outwitting the blood-brain barrier for therapeutic purposes: Osmotic opening and other means. *Neurosurgery.* 42(5):1083.

109. Doolittle ND, Abrey LE, Ferrari N, Hall WA, Laws ER, McLendon RE, Muldoon LL et al., 2002. Targeted delivery in primary and metastatic brain tumors: Summary report of the seventh annual meeting of the Blood-Brain Barrier Disruption Consortium. *Clin Cancer Res.* 8(6):1702.

110. Hynynen K, McDannold N, Vykhodtseva N, Jolesz FA, 2001. Noninvasive MR imaging-guided focal opening of the blood-brain barrier in rabbits. *Radiology.* 220(3):640.

111. Hynynen K, McDannold N, Vykhodtseva N, Raymond S, Weissleder R, Jolesz FA, Sheikov N, 2006. Focal disruption of the blood-brain barrier due to 260-kHz ultrasound bursts: A method for molecular imaging and targeted drug delivery. *J Neurosurg.* 105(3):445.

112. Kinoshita M, 2006. Targeted drug delivery to the brain using focused ultrasound. *Top Magn Reson Imaging*. 17(3):209.

113. Dean RL, Emerich DF, Hasler BP, Bartus RT, 1999. Cereport (RMP-7) increases carboplatin levels in brain tumors after pretreatment with dexamethasone. *Neuro Oncol*. 1(4):268.

114. Borlongan CV, Emerich DF, 2003. Facilitation of drug entry into the CNS via transient permeation of blood brain barrier: Laboratory and preliminary clinical evidence from bradykinin receptor agonist, *Cereport*. *Brain Res Bull*. 60(3):297.

115. Gabathuler R, 2010. Approaches to transport therapeutic drugs across the blood-brain barrier to treat brain diseases. *Neurobiol Dis*. 37(1):48.

116. Lipinski CA, 2000. Drug-like properties and the causes of poor solubility and poor permeability. *J Pharmacol Toxicol Methods*. 44(1):235.

117. Shashoua VE, Hesse GW, 1996. N-docosahexaenoyl, 3 hydroxytyramine: A dopaminergic compound that penetrates the blood-brain barrier and suppresses appetite. *Life Sci*. 58(16):1347.

118. Sawada GA, Williams LR, Lutzke BS, Raub TJ, 1999. Novel, highly lipophilic antioxidants readily diffuse across the blood-brain barrier and access intracellular sites. *J Pharmacol Exp Ther*. 288(3):1327.

119. Bodor N, Farag HH, Brewster ME 3rd, 1981. Site-specific, sustained release of drugs to the brain. *Science* 214(4527):1370.

120. Brewster ME, Raghavan K, Pop E, Bodor N, 1994. Enhanced delivery of ganciclovir to the brain through the use of redox targeting. *Antimicrob Agents Chemother* 38(4):817.

121. Pardridge WM, 2003. Blood-brain barrier drug targeting: The future of brain drug development. *Mol Interv*. 3(2):90.

122. Pardridge WM, 2006. Molecular Trojan horses for blood-brain barrier drug delivery. *Curr. Opin. Pharm*. 6(5):494.

123. Miller G, 2002. Drug targeting. Breaking down barriers. *Science* 297(5584):1116.

124. Gosk S, Vermehren C, Storm G, Moos T, 2004. Targeting anti-transferrin receptor antibody (OX26) and OX26-conjugated liposomes to brain capillary endothelial cells using *in situ* perfusion. *J Cereb Blood Flow Metab*. 24(11):1193.

125. Kumagai AK, Eisenberg JB, Pardridge WM, 1987. Absorptive-mediated endocytosis of cationized albumin and a beta-endorphin-cationized albumin chimeric peptide by isolated brain capillaries. Model system of blood-brain barrier transport. *J Biol Chem*. 262(31):15214.

126. Lawrenson JG, Reid AR, Allt G, 1997. Molecular characteristics of pial microvessels of the rat optic nerve. Can pial microvessels be used as a model for the blood-brain barrier? *Cell Tissue Res*. 288, 259–65.

127. Yuan W, Li G, Zeng M, Fu BM, 2010. Modulation of the blood-brain barrier permeability by plasma glycoprotein orosomucoid. *Microvasc. Res*. 80(1):148–157.

128. Hervé F, Ghinea N, Scherrmann JM, 2008. CNS delivery via adsorptive transcytosis. *AAPS J*. 10(3):455.

129. Rousselle C, Clair P, Smirnova M, Kolesnikov Y, Pasternak GW, Gac-Breton S, Rees AR, Scherrmann JM, Temsamani J, 2003. Improved brain uptake and pharmacological activity of dalargin using a peptide-vector-mediated strategy. *J Pharmacol Exp Ther*. 306(1):371.

130. Liu L, Guo K, Lu J, Venkatraman SS, Luo D, Ng KC, Ling EA, Moochhala S, Yang YY, 2008. Biologically active core/shell nanoparticles self-assembled from cholesterol-terminated PEG-TAT for drug delivery across the blood-brain barrier. *Biomaterials* 29(10):1509.

131. Lockman PR, Koziara JM, Mumper RJ, Allen DD, 2004. Nanoparticle surface charges alter blood-brain barrier integrity and permeability. *J Drug Target*. 12(9–10):635.

132. Szmydynger-Chodobska J, Chodobski A, Johanson CE, 1994. Postnatal developmental changes in blood flow to choroid plexuses and cerebral cortex of the rat. *Am J Physiol*. 266(5 Pt 2):R1488.

133. Kramer K, Humm JL, Souweidane MM, Zanzonico PB, Dunkel IJ, Gerald WL, Khakoo Y et al., 2007. Phase I study of targeted radioimmunotherapy for leptomeningeal cancers using intra-Ommaya 131-I-3F8. *J Clin Oncol*. 25:5465–5470.

134. Bruno MK, Raizer J, 2005. Leptomeningeal metastases from solid tumors (meningeal carcinomatosis). *Cancer Treat Res.* 125:31.

135. Gleissner B, Chamberlain MC, 2006. Neoplastic meningitis. *Lancet Neurol.* 5:443.

136. Grossman SA, Spence A, 1999. NCCN clinical practice guidelines for carcinomatous/lymphotous meningitis. *Oncology.* 13:144.

137. Kramer K, Kushner B, Heller G, Cheung NK, 2001. Neuroblastoma metastatic to the central nervous system. The Memorial Sloan-Kettering Cancer Center Experience and a Literature Review. *Cancer.* 91:1510.

138. Lv Y, Cheung NK, Fu BM, 2009. A pharmacokinetic model for radioimmunotherapy delivered through cerebrospinal fluid for the treatment of leptomeningeal metastases. *J Nuclear Med.* 50(8):1324.

9

Interstitial Transport in the Brain: Principles for Local Drug Delivery

9.1   Introduction **9**-1
9.2   Implantable Controlled Delivery Systems for Chemotherapy .. **9**-2
9.3   Drug Transport after Release from the Implant **9**-4
9.4   Application of Diffusion–Elimination Models to Intracranial BCNU Delivery Systems **9**-7
9.5   Limitations and Extensions of the Diffusion–Elimination Model **9**-9
     Failure of the Model in Certain Situations
9.6   New Approaches to Drug Delivery Suggested by Modeling.... **9**-12
9.7   Conclusion **9**-12
References **9**-12

W. Mark Saltzman
Yale University

9.1 Introduction

Traditional methods for delivering drugs to the brain are inadequate. Many drugs, particularly water-soluble or high-molecular-weight compounds, do not enter the brain following systemic administration because they permeate through blood capillaries very slowly. This blood–brain barrier (BBB) severely limits the number of drugs that are candidates for treating brain disease.

Several strategies for increasing the permeability of brain capillaries to drugs have been tested. Since the BBB is generally permeable to lipid-soluble compounds that can dissolve and diffuse through endothelial cell membranes [1,2], a common approach for enhancing brain delivery of compounds is chemical modification to enhance lipid solubility [3]. Unfortunately, lipidization approaches are not useful for drugs with molecular weight larger than 1000. Another approach for increasing permeability is the entrapment of drugs in liposomes [4], but delivery may be limited by liposome stability in the plasma and uptake at other tissue sites.

Specific nutrient transport systems in brain capillaries can be used to facilitate drug entry into the brain. L-Dopa (L-3,4-dihydroxyphenylalanine), a metabolic precursor of dopamine, is transported across endothelial cells by the neutral amino acid transport system [5]. L-Dopa permeates through capillaries into the striatal tissue, where it is decarboxylated to form dopamine. Therefore, systemic administration of L-dopa is often beneficial to patients with Parkinson's disease. Certain protein modifications, such as cationization [6] and anionization [7], produce enhanced uptake in the brain. Modification of drugs [8,9] by linkage to an antitransferrin receptor antibody also appears to enhance transport into the brain. This approach depends on receptor-mediated transcytosis of transferrin–receptor complexes by brain endothelial cells; substantial uptake also occurs in the liver.

The permeability of brain capillaries can be transiently increased by intra-arterial injection of the hyperosmolar solutions, which disrupt interendothelial tight junctions [10]. But BBB disruption affects capillary permeability throughout the brain, enhancing permeability to all compounds in the blood, not just the agent of interest. Intraventricular therapy, where agents are administered directly into the cerebrospinal fluid (CSF) of the ventricles, results in high concentrations within the brain tissue, but only in regions immediately surrounding the ventricles [11,12]. Because the agent must diffuse into the brain parenchyma from the ventricles, and because of the high rate of clearance of agents in the CNS into the peripheral circulation, this strategy cannot be used to deliver agents deep into the brain.

Because of the difficulty in achieving therapeutic drug levels by systemic administration, methods for direct administration of drugs into the brain parenchyma have been tested. Drugs can be delivered directly into the brain tissue by infusion, implantation of a drug-releasing matrix, or transplantation of drug-secreting cells [13]. These approaches provide sustained drug delivery that can be limited to specific sites, localizing therapy to a brain region. Because these methods provide a localized and continuous source of active drug molecules, the total drug dose can be less than needed with systemic administration. With polymeric controlled release, the implants can also be designed to protect unreleased drug from degradation in the body and to permit localization of extremely high doses (up to the solubility of the drug) at precisely defined locations in the brain. Infusion systems require periodic refilling; the drug is usually stored in a liquid reservoir at body temperature and many drugs are not stable under these conditions.

This chapter describes the transport of drug molecules that are directly delivered into the brain. For purposes of clarity, a specific example is considered: polymeric implants that provide controlled release of chemotherapy. The results can be extended to other modes of administration [13,14] and other types of drug agents [15].

9.2 Implantable Controlled Delivery Systems for Chemotherapy

The kinetics of drug release from a controlled release system are usually characterized *in vitro*, by measuring the amount of drug released from the matrix into a well-stirred reservoir of phosphate buffered water or saline at 37°C. Controlled release profiles for some representative anticancer agents are shown in Figure 9.1; all of the agents selected for these studies—1,3-bis(2-chloroethyl)-1-nitrosourea (BCNU), 4-HC, cisplatin, and taxol—are used clinically for chemotherapy of brain tumors. The controlled release period can vary from several days to many months, depending on properties of the drug, the polymer, and the method of formulation. Therefore, the delivery system can be tailored to the therapeutic situation by manipulation of implant properties.

The release of drug molecules from polymer matrices can be regulated by diffusion of drug through the polymer matrix or degradation of the polymer matrix. In many cases (including the release of BCNU, cisplatin, and 4HC from the degradable matrices shown in Figure 9.1), drug release from biodegradable polymers appears to be diffusion-mediated, probably because the time for polymer degradation is longer than the time required for drug diffusion through the polymer. In certain cases, linear release, which appears to correlate with the polymer degradation rate, can be achieved; this might be the case for taxol release from the biodegradable matrix (Figure 9.1), although the exceedingly low solubility of taxol in water may also contribute substantially to the slowness of release.

For diffusion-mediated release, the amount of drug released from the polymer is proportional to the concentration gradient of the drug in the polymer. By performing a mass balance on drug molecules within a differential volume element in the polymer matrix, a conservation equation for drug within the matrix is obtained:

$$\frac{\partial C_p}{\partial t} = D_p \nabla^2 C_p$$

(9.1)

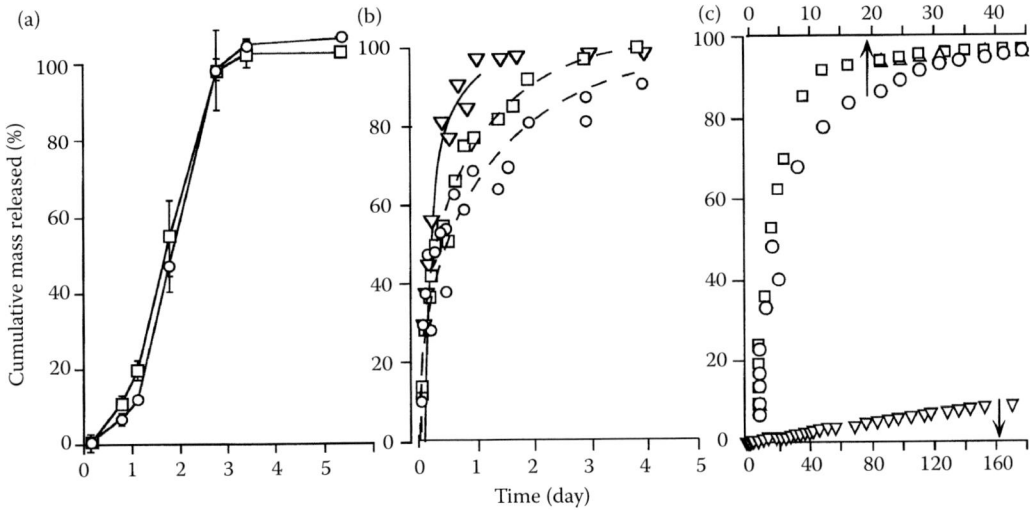

FIGURE 9.1 Controlled release of anticancer compounds from polymeric matrices. (a) Release of cisplatin (circles) from a biodegradable copolymer of fatty acid dimers and sebacic acid, p(FAD:SA), initially containing 10% drug. (b) Release of BCNU from EVAc matrices (circles), polyanhydride matrices p(CPP:SA) (squares), and p(FAD:SA) (triangles) matrices initially containing 20% drug. (c) Release of BCNU (squares), 4HC (circles), and taxol (triangles) from p(CPP:SA) matrices initially containing 20% drug. Note that panel (c) has two time axes: the lower axis applies to the release of taxol and the upper axis applies to the release of BCNU and 4HC. (Adapted from Dang, W. and W.M. Saltzman, *Journal of Biomaterials Science, Polymer Edition*, 1994. 6(3):291–311; Dang, W., *Engineering Drugs and Delivery Systems for Brain Tumor Therapy*. 1993, The Johns Hopkins University: Baltimore, MD.)

where C_p is the local concentration of drug in the polymer and D_p is the diffusion coefficient of the drug in the polymer matrix. This equation can be solved, with appropriate boundary and initial conditions, to obtain the cumulative mass of drug released as a function of time; the details of this procedure are described elsewhere [18]. A useful approximate solution, which is valid for the initial 60% of release, is

$$M_t = 4M_o\sqrt{\frac{D_{i:p}t}{\pi L^2}} \tag{9.2}$$

where M_t is the cumulative mass of drug released from the matrix, M_o is the initial mass of drug in the matrix, and L is the thickness of the implant. By comparing Equation 9.2 to the experimentally determined profiles, the rate of diffusion of the agent in the polymer matrix can be estimated (Table 9.1).

TABLE 9.1 Diffusion Coefficients for Chemotherapy Drug Release from Biocompatible Polymer Matrices

Drug	Polymer	Initial Loading (%)	D_p (cm^2/s)
Cisplatin	P(FAD:SA)	10	6.8×10^{-9}
BCNU	EVAc	20	1.6×10^{-8}
BCNU	P(FAD:SA)	20	6.9×10^{-8}
BCNU	P(CPP:SA)	20	2.3×10^{-8} (panel b)
			2.0×10^{-8} (panel c)
4HC	P(CPP:SA)	20	3.1×10^{-10}
Taxol	P(CPP:SA)	20	n.a.

Note: Diffusion coefficients were obtained by comparing the experimental data shown in Figure 9.1 to Equation 9.2 and determining the best value of the diffusion coefficient to represent the data. n.a., not applicable.

9.3 Drug Transport after Release from the Implant

Bypassing the BBB is necessary, but not sufficient for effective drug delivery. Consider the consequences of implanting a delivery system, such as one of the materials characterized above, within the brain. Molecules released into the interstitial fluid in the brain extracellular space (ECS) must penetrate into the brain tissue to reach tumor cells distant from the implanted device. Before these drug molecules can reach the target site, however, they might be eliminated from the interstitium by partitioning into brain capillaries or cells, entering the CSF, or being inactivated by extracellular enzymes. Elimination always accompanies dispersion; therefore, regardless of the design of the delivery system, one must understand the dynamics of both processes in order to predict the spatial pattern of drug distribution after delivery.

The polymer implant is surrounded by biological tissue, composed of cells and an ECS filled with extracellular fluid (ECF). Immediately following implantation, drug molecules escape from the polymer and penetrate the tissue. Once in the brain tissue, drug molecules: (1) diffuse through the tortuous ECS in the tissue; (2) diffuse across semipermeable tissue capillaries to enter the systemic circulation and, therefore, are removed from the brain tissue; (3) diffuse across cell membranes by passive, active, or facilitated transport paths, to enter the intracellular space; (4) transform, spontaneously or by an enzyme-mediated pathway, into other compounds; and (5) bind to fixed elements in the tissue. Each of these events influence drug therapy: diffusion through the ECS is the primary mechanism of drug distribution in brain tissue; elimination of the drug occurs when it is removed from the ECF or transformed, and binding or internalization may slow the progress of the drug through the tissue.

A mass balance on a differential volume element in the tissue [19] gives a general equation describing drug transport in the region near the polymer [20]:

$$\frac{\partial C_i}{\partial t} + \bar{v} \cdot \nabla C_t = D_b \nabla^2 C_i + R_e(C_i) - \frac{\partial B}{\partial t} \tag{9.3}$$

where C is the concentration of the diffusible drug in the tissue surrounding the implant (g/cm^3 tissue), v is the fluid velocity (cm/s), D_b is the diffusion coefficient of the drug in the tissue (cm^2/s), $R_e(C)$ is the rate of drug elimination from the ECF (g/s-cm^3 tissue), B is the concentration of drug bound or internalized in cells (g/cm^3 tissue), and t is the time following implantation. In deriving this equation, the conventions developed by Nicholson [21], based on volume-averaging in a complex medium, and Patlak and Fenstermacher [20] were combined. In this version of the equation, the concentrations C and B and the elimination rate $R_e(C)$ are defined per unit volume of tissue. D_b is an effective diffusion coefficient, which must be corrected from the diffusion coefficient for the drug in water to account for the tortuosity of the ECS.

When the binding reactions are rapid, the amount of intracellular or bound drug can be assumed to be directly proportional, with an equilibrium coefficient K_{bind}, to the amount of drug available for internalization or binding:

$$B = K_{bind}C \tag{9.4}$$

Substitution of Equation 9.4 into Equation 9.3 yields, with some simplification:

$$\frac{\partial C_t}{\partial t} = \frac{1}{1 + K_{bind}}(D_b \nabla^2 C_t + R_e(C_t) - \bar{v} \cdot \nabla C_t) \tag{9.5}$$

The drug elimination rate, $R_e(C)$, can be expanded into the following terms:

$$R_e(C_t) = k_{bbb}\left(\frac{C_t}{\varepsilon_{ecs}} - C_{plasma}\right) + \frac{V_{max}C_t}{K_m + C_t} + k_{ne}C_t \qquad (9.6)$$

where k_{bbb} is the permeability of the BBB (defined based on concentration in the ECS), C_{plasma} is the concentration of drug in the blood plasma, V_{max} and K_m are Michaelis–Menton constants, and k_{ne} is a first-order rate constant for drug elimination due to nonenzymatic reactions. For any particular drug, some of the rate constants may be very small, reflecting the relative importance of each mechanism of drug elimination. If it is assumed that the permeability of the BBB is low ($C_{pl} \ll C$) and the concentration of drug in the brain is sufficiently low so that any enzymatic reactions are in the first-order regime ($C \ll K_m$), Equation 9.6 can be reduced to

$$-R_e(C_t) = \frac{k_{bbb}}{\varepsilon_{ecs}}C_t + \frac{V_{max}}{K_m}C_t + k_{ne}C_t = k_{app}C_t \qquad (9.7)$$

where k_{app} is a lumped first-order rate constant. With these assumptions, Equation 9.4 can be simplified by definition of an apparent diffusion coefficient, D^*, and an apparent first-order elimination constant, k^*:

$$\frac{\partial C_t}{\partial t} = D^*\nabla^2 C_t + k^*C_t - \frac{\bar{v} \cdot \nabla C_t}{1 + K_{bind}} \qquad (9.8)$$

where $k^* = k_{app}/(1 + K_{bind})$ and $D = D_b/(1 + K_{bind})$.

Boundary and initial conditions are required for solution of differential Equation 9.8. If a spherical implant of radius R is implanted into a homogeneous region of the brain, at a site sufficiently far from anatomical boundaries, the following assumptions are reasonable:

$$C_t = 0 \quad \text{for } t = 0; \quad r > R \qquad (9.9)$$

$$C_t = C_i \quad \text{for } t > 0; \quad r = R \qquad (9.10)$$

$$C_t = 0 \quad \text{for } t > 0; \quad r \to \infty \qquad (9.11)$$

In many situations, drug transport due to bulk flow can be neglected. This assumption (v is zero) is common in previous studies of drug distribution in brain tissue [20]. For example, in a previous study of cisplatin distribution following continuous infusion into the brain, the effects of bulk flow were found to be small, except within 0.5 mm of the site of infusion [22]. In the cases considered here, since drug molecules enter the tissue by diffusion from the polymer implant, not by pressure-driven flow of a fluid, no flow should be introduced by the presence of the polymer. With fluid convection assumed negligible, the general governing equation in the tissue, Equation 9.8, reduces to

$$\frac{\partial C_i}{\partial t} = D^*\nabla^2 C_t + k^*C_t \qquad (9.12)$$

The no-flow assumption may be inappropriate in certain situations. In brain tumors, edema and fluid movement are significant components of the disease. In addition, some drugs can elicit cytotoxic edema. Certain drug/polymer combinations can also release drugs in sufficient quantity to create density-induced fluid convection.

Equation 9.12, with conditions 9.9 through 9.11, can be solved by Laplace transform techniques [13] to yield

$$\frac{C_t}{C_i} = \frac{1}{2\zeta}\left\{\exp[-\phi(\zeta-1)]\mathrm{erfc}\left[\frac{\zeta-1}{2\sqrt{\tau}} - \phi\sqrt{\tau}\right] + \exp[\phi(\zeta-1)]\mathrm{erfc}\left[\frac{\zeta-1}{2\sqrt{\tau}} + \phi\sqrt{\tau}\right]\right\} \qquad (9.13)$$

where the dimensionless variables are defined as follows:

$$\zeta = \frac{r}{R}; \quad \tau = \frac{D^*t}{R^2}; \quad \phi = R\sqrt{\frac{k^*}{D^*}} \qquad (9.14)$$

The differential equation also has a steady-state solution, which is obtained by solving Equation 9.12 with the time derivative set equal to zero and subject to the boundary conditions 9.9 and 9.10:

$$\frac{C_t}{C_i} = \frac{1}{\zeta}\exp[-\phi(\zeta-1)] \qquad (9.15)$$

Figure 9.2 shows concentration profiles calculated using Equations 9.13 and 9.15. In this situation, which was obtained using reasonable values for all of the parameters, steady state is reached approximately 1 h after implantation of the delivery device. The time required to achieve steady state depends

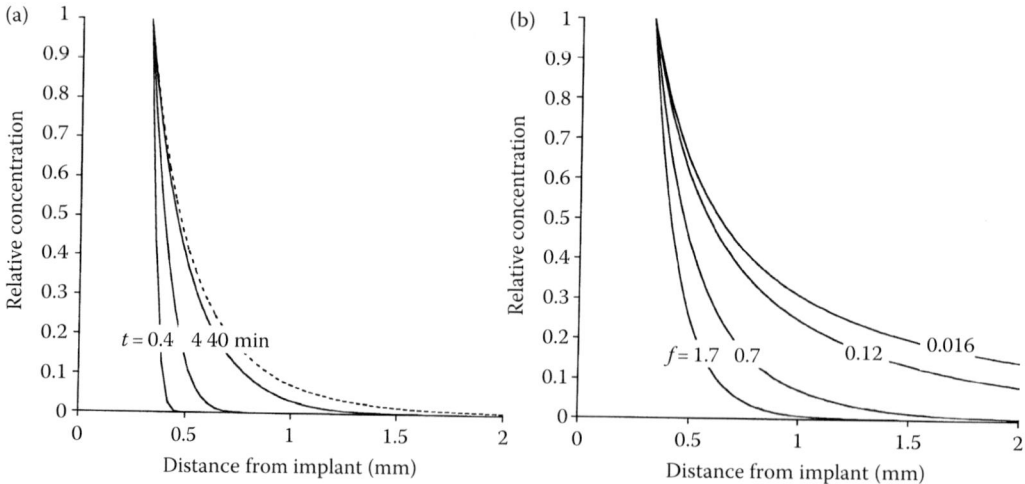

FIGURE 9.2 Concentration profiles after implantation of a spherical drug-releasing implant. (Panel a, transient) Solid lines represent the transient solution to Equation 9.12 (i.e., Equation 9.13) with the following parameter values: $D^* = 4 \times 10^{-7}$ cm²/s; $R = 0.032$ cm; $k^* = 1.9 \times 10^{-4}$ s⁻¹ ($t^{1/2} = 1$ h). The dashed line represents the steady-state solution (i.e., Equation 9.15) for the same parameters. (Panel b, steady state) Solid lines in this plot represent Equation 9.15 with the following parameters: $D^* = 4 \times 10^{-7}$ cm²/s; $R = 0.032$ cm. Each curve represents the steady-state concentration profile for drugs with different elimination half-lives in the brain, corresponding to different dimensionless moduli, f: $t^{1/2} = 10$ min ($f = 1.7$); 1 h (0.7); 34 h (0.12); and 190 h (0.016).

on the rate of diffusion and elimination, as previously described [23], but will be significantly <24 h for most drug molecules.

9.4 Application of Diffusion–Elimination Models to Intracranial BCNU Delivery Systems

The preceding mathematical analysis, which assumes diffusion and first-order elimination in the tissue, agrees well with experimental concentration profiles obtained after implantation of controlled release polymers (Figure 9.3). At 3, 7, and 14 days after implantation of a BCNU-releasing implant, the concentration profile at the site of the implant was very similar. The parameter values (obtained by fitting Equation 9.15 to the experimental data) were consistent with parameters obtained using other methods [24], suggesting that diffusion and first-order elimination were sufficient to account for the pattern of drug concentration observed during this period. Parameter values were similar at 3, 7, and 14 days, indicating that the rates of drug release, drug dispersion, and drug elimination did not change during this period. This equation has been compared to concentration profiles measured for a variety of molecules delivered by polymer implants to the brain—dexamethasone [23], molecular weight fractions of dextran [25], nerve growth factor in rats [26,27], BCNU in rats [24], rabbits [28], and monkeys [29]. In each of these cases, the steady-state diffusion–elimination model appears to capture most of the important features of drug transport.

This model can be used to develop guidelines for the design of intracranial delivery systems. Table 9.2 lists some of the important physical and biological characteristics of a few compounds that have been considered for interstitial delivery to treat brain tumors. When the implant is surrounded by tissue, the

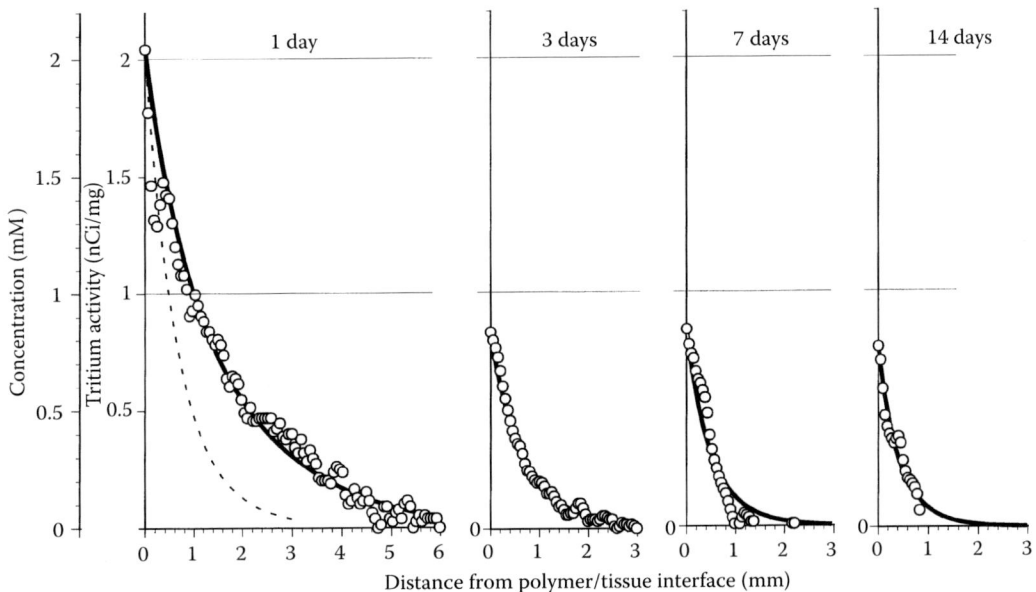

FIGURE 9.3 Concentration profiles after implantation of a BCNU-releasing implant. Solutions to Equation 9.15 were compared to experimental data obtained by quantitative autoradiographic techniques. The solid lines in the three panels labeled 3, 7, and 14 days were all obtained using the following parameters: $R = 0.15$ cm; $f = 2.1$; and $C_I = 0.81$ mM. The solid line in the panel labeled 1 day was obtained using the following parameters: $R = 0.15$ cm; $f = 0.7$; and $C_I = 1.9$ mM. (Modified from Fung, L. et al., *Pharmaceutical Research*, 1996. 13:671–682.)

TABLE 9.2   Implant Design Applied to Three Chemotherapy Compounds

	BCNU	4HC	Methotrexate
Molecular weight (daltons)	214	293	454
Solubility (mM)	12	100	100
$\text{Log}_{10}K$	1.53	0.6	−1.85
k^* (h)	70	70	1
D^* (10^{-7} cm²/s)	14	14	5
Toxic concentration in culture (µM)	25	10	0.04
Maximum release rate (mg/day)	1.2	14	17
Implant lifetime at max rate (day)	0.85	0.07	0.06
Maximum concentration in tissue for 1-week-releasing implant (mM)	1.5	1.1	1.8
RT (mm)	1.3	2.5	5

Note: *K* is the octanol:water partition coefficient, *k* is the rate of elimination due to permeation through capillaries, and D_b is the diffusion coefficient of the drug in the brain. The following values are assumed, consistent with our results from polymer delivery to rats and rabbits: radius of spherical implant $R = 1.5$ mm; mass of implant $M = 10$ mg; drug loading in implant load = 10%.

maximum rate of drug release is determined by the solubility of the drug, C_s, and the rate of diffusive transport through the tissue:

$$\left(\frac{dM_t}{dt}\right)_{\max} = (\text{Maximum flux}) \times (\text{Surface area}) = -D^* \left.\frac{\partial C_t}{\partial r}\right|_R 4\pi R^2 \tag{9.16}$$

Evaluating the derivative in Equation 9.16 from the steady-state concentration profile (Equation 9.15) yields

$$\left(\frac{dM_i}{dt}\right)_{\max} = 8\pi \dot{D}^* C_s R \tag{9.17}$$

Regardless of the properties of the implant, it is not possible to release drug into the tissue at a rate faster than determined by Equation 9.17. If the release rate from the implant is less than this maximum rate, C_i (the concentration in the tissue immediately outside the implant) is less than the saturation concentration, C_s. The actual concentration C_i can be determined by balancing the release rate from the implant (dM_t/dt, which can be determined from Equation 9.2 provided that diffusion is the mechanism of release from the implant) with the rate of penetration into the tissue obtained by substituting C_i for C_s in Equation 9.17:

$$C^* = \frac{dM_i}{dt}\left(\frac{1}{8\pi D^* R}\right) \tag{9.18}$$

The effective region of therapy can be determined by calculating the distance from the surface of the implant to the point where the concentration drops below the cytotoxic level (estimated as the cytotoxic concentration determined from *in vitro* experiments). Using Equation 9.15, and defining the radial distance for effective treatment as R_T, yields

$$\frac{C_{\text{cytotoxic}}}{C_i} = \frac{R}{R_T}\exp\left\{-R\sqrt{\frac{k^*}{D^*}}\left(\frac{R_T}{R}-1\right)\right\} \tag{9.19}$$

Alternately, an effective penetration distance, d_p, can be defined as the radial position at which the drug concentration has dropped to 10% of the peak concentration:

$$0.10 = \frac{R}{d_p} \exp\left\{-R\sqrt{\frac{k^*}{D^*}}\left(\frac{d_p}{R} - 1\right)\right\} \tag{9.20}$$

These equations provide quantitative criteria for evaluating the suitability of chemotherapy agents for direct intracranial delivery (Table 9.2).

9.5 Limitations and Extensions of the Diffusion–Elimination Model

9.5.1 Failure of the Model in Certain Situations

The previous section outlined one method for the analysis of drug transport after implantation of a drug-releasing device. A simple pseudo-steady-state equation (Equation 9.15) yielded simple guidelines (Equations 9.16 through 9.19) for device design. Because the assumptions of the model were satisfied over a substantial fraction of the release period (days 3–14, based on the data shown in Figure 9.3), this analysis may be useful for predicting the effects of BCNU release from biodegradable implants. Pseudo-steady-state assumptions are reasonable during this period of drug release, presumably because the time required to achieve steady state (which is on the order of minutes) is much less than the characteristic time associated with changes in the rate of BCNU release from the implant (which is on the order of days).

But experimental concentration profiles measured 1 day after implantation were noticeably different: the peak concentration was substantially higher and the drug penetration into the surrounding tissue was deeper (see the left-hand panel of Figure 9.3). This behavior cannot be easily explained by the pseudo-steady-state models described above. For example, if the difference observed at 1 day represents transient behavior, the concentration observed at a fixed radial position should increase with time (Figure 9.2); in contrast, the experimental concentration at any radial position on day 1 is higher than the concentration measured at that same position on subsequent days.

9.5.1.1 Effect of Drug Release Rate

Alternately, the observed difference at 1 day might be due to variability in the rate of BCNU release from the polymer implant over this period, with transport characteristics in the tissue remaining constant. When similar BCNU-releasing implants are tested *in vitro*, the rate of drug release decreased over time (Figure 9.1). Equation 9.18 predicts the variation in peak concentration with release rate; the twofold higher concentration observed at the interface on day 1 (as compared to days 3 through 14) could be explained by a twofold higher release rate on day 1. But the effective penetration distance, d_p, does not depend on release rate. Experimentally measured penetration distances are ~1.4 mm on days 3, 7, and 14 and ~5 mm on day 1. This difference in penetration is shown more clearly in the day 1 panel of Figure 9.3: the dashed line shows the predicted concentration profile if k^* and D^* were assumed equal to the values obtained for days 3, 7, and 14. Changes in the rate of BCNU release are insufficient to explain the differences observed experimentally.

9.5.1.2 Determinants of Tissue Penetration

Penetration of BCNU is enhanced at day 1 relative to penetration at days 3, 7, and 14. For an implant of fixed size, penetration depends only on the ratio of elimination rate to diffusion rate: k^*/D^*. Increased penetration results from a decrease in this ratio (Figure 9.2), which could occur because of either a decreased rate of elimination (smaller k^*) or an increased rate of diffusion (larger D^*). But there are no good reasons to believe that BCNU diffusion or elimination are different on day 1 than on days 3 through 14. With its high lipid solubility, BCNU can diffuse readily through brain tissue. In addition, elimination of BCNU from the brain occurs predominantly by partitioning into the circulation; since BCNU can permeate the capillary wall by diffusion, elimination is not a saturable process. Perhaps the enhanced penetration of BCNU is due to the presence of another process for drug dispersion, such as bulk fluid flow, which was neglected in the previous analysis.

The diffusion/elimination model compares favorably with available experimental data, but the assumptions used in predicting concentration profiles in the brain may not be appropriate in all cases. Deviations from the predicted concentration profiles may occur due to extracellular fluid flows in the brain, complicated patterns of drug binding, or multistep elimination pathways. The motion of interstitial fluid in the vicinity of the polymer and the tumor periphery may not always be negligible, as mentioned above. The interstitial fluid velocity is proportional to the pressure gradient in the interstitium; higher interstitial pressure in tumors—due to tumor cell proliferation, high vascular permeability, and the absence of functioning lymphatic vessels—may lead to steep interstitial pressure gradients at the periphery of the tumor [30]. As a result, interstitial fluid flows within the tumor may influence drug transport. A drug at the periphery of the tumor must overcome outward convection to diffuse into the tumor. Similarly, local edema after surgical implantation of the polymer may cause significant fluid movement in the vicinity of the polymer. More complete mathematical models that include the convective contribution to drug transport are required.

9.5.1.3 Effect of Fluid Convection

When bulk fluid flow is present ($v \neq 0$), concentration profiles can be predicted from Equation 9.8, subject to the same boundary and initial conditions (Equations 9.9 through 9.11). In addition to Equation 9.8, continuity equations for water are needed to determine the variation of fluid velocity in the radial direction. This set of equations has been used to describe concentration profiles during microinfusion of drugs into the brain [14]. Relative concentrations were predicted by assuming that the brain behaves as a porous medium (i.e., velocity is related to pressure gradient by Darcy's law). Water introduced into the brain can expand the interstitial space; this effect is balanced by the flow of water in the radial direction away from the infusion source and, to a lesser extent, by the movement of water across the capillary wall.

In the presence of fluid flow, penetration of drug away from the source is enhanced (Figure 9.4). The extent of penetration depends on the velocity of the flow and the rate of elimination of the drug. These calculations were performed for macromolecular drugs, which have limited permeability across the brain capillary wall. The curves indicate steady-state concentration profiles for three different proteins with metabolic half-lives of 10 min, 1 h, and 33.5 h. In the absence of fluid flow, drugs with longer half-lives penetrate deeper into the tissue (solid lines in Figure 9.4 were obtained from Equation 9.15). This effect is amplified by the presence of flow (dashed lines in Figure 9.4).

During microinfusion, drug is introduced by pressure-driven fluid flow from a small catheter. Therefore, pressure gradients are produced in the brain interstitial space, which lead to fluid flow through the porous brain microenvironment. Volumetric infusion rates of 3 μL/min were assumed in the calculations reproduced in Figure 9.4. Since loss of water through the brain vasculature is slight, the velocity can be determined as a function of radial position:

$$v_r = \frac{q}{4\pi r^2 \varepsilon} \tag{9.21}$$

where q is the volumetric infusion rate and ε is the volume fraction of the interstitial space in the brain (approximately 0.20). Fluid velocity decreases with radial distance from the implant (Table 9.3), but at all locations within the first 20 mm of the implant site, predicted velocity was much greater than the velocities reported previously during edema or tumor growth in the brain.

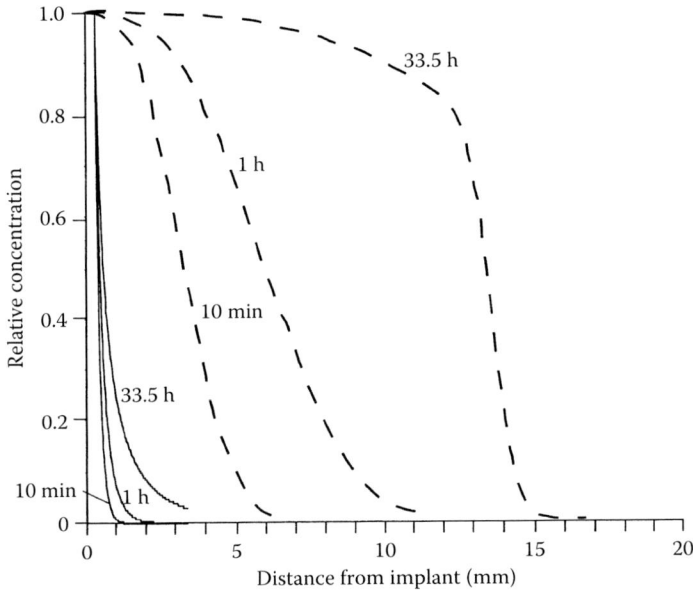

FIGURE 9.4 Concentration profiles predicted in the absence (solid lines) and presence (dashed lines) of interstitial fluid flow. Solid lines were obtained from Equation 9.15 with the following parameter values: $R = 0.032$ cm; $D = 4 \times 10^{-7}$ cm²/s; and $k^* = \ln(2)/t^{1/2}$, where $t^{1/2}$ is 10 min, 1 h, or 33.5 h as indicated on the graph. Dashed lines were obtained from Reference 14 using the same parameter values and an infusion rate of 3 µL/min. The dashed line indicating the interstitial flow calculation for the long-lived drug ($t^{1/2} = 33.5$ h) was not at steady state, but at 12 h after initiation of the flow.

TABLE 9.3 Interstitial Fluid Velocity as a Function of Radial Position during Microinfusion

Radial Position (mm)	Interstitial Velocity (µm/s)
2	5.0
5	0.8
10	0.2
20	0.05

Source: Calculated by method reported in Morrison, P.F. et al., *American Journal of Physiology*, 1994. 266:R292–R305.

The profiles predicted in Figure 9.4 were associated with the introduction of substantial volumes of fluid at the delivery site. Flow-related phenomena are probably less important in drug delivery by polymer implants. Still, this model provides useful guidelines for predicting the influence of fluid flow on local rates of drug movement. Clearly, the effect of flow velocity on drug distribution is substantial (Figure 9.4). Even relatively low flows, perhaps as small as 0.03 µm/s, are large enough to account for the enhancement in BCNU penetration observed at day 1 in Figure 9.3.

9.5.1.4 Effect of Metabolism

The metabolism, elimination, and binding of drug are assumed to be first-order processes in our simple analysis. This assumption may not be realistic in all cases, especially for complex agents such as proteins. The metabolism of drugs in normal and tumor tissues is incompletely understood. Other cellular factors (e.g., the heterogeneity of tumor-associated antigen expression and multidrug resistance) that influence the uptake of therapeutic agents may not be accounted for by our simple first-order elimination.

Finally, changes in the brain that occur during the course of therapy are not properly considered in this model. Irradiation can be safely administered when a BCNU-loaded polymer has been implanted in monkey brains, suggesting the feasibility of adjuvant radiotherapy. However, irradiation also causes necrosis in the brain. The necrotic region has a lower perfusion rate and interstitial pressure than tumor tissue; thus, the convective interstitial flow due to fluid leakage is expected to be smaller. Interstitial diffusion of macromolecules is higher in tumor tissue than in normal tissue, as the tumor tissue has larger interstitial spaces [31]. The progressive changes in tissue properties—due to changes in tumor size, irradiation, and activity of chemotherapy agent—may be an important determinant of drug transport and effectiveness of therapy in the clinical situation.

9.6 New Approaches to Drug Delivery Suggested by Modeling

Mathematical models, which describe the transport of drug following controlled delivery, can predict the penetration distance of drug and the local concentration of drug as a function of time and location. The calculations indicate that drugs with slow elimination will penetrate deeper into the tissue. The modulus f, which represents the ratio of elimination to diffusion rates in the tissue, provides a quantitative criterion for selecting agents for interstitial delivery. For example, high-molecular-weight dextrans were retained longer in the brain space, and penetrated a larger region of the brain than low-molecular-weight molecules following release from an intracranial implant [25]. This suggests a strategy for modifying molecules to improve their tissue penetration by conjugating active drug molecules to inert polymeric carriers. For conjugated drugs, the extent of penetration should depend on the modulus f for the conjugated compound as well as the degree of stability of the drug–carrier linkage.

The effects of conjugation and stability of the linkage between drug and carrier on enhancing tissue penetration in the brain have been studied in a model system [32]. Methotrexate (MTX)–dextran conjugates with different dissociation rates were produced by linking MTX to dextran (molecular weight 70,000) through a short-lived ester bond (half-life ≈3 days) and a longer-lived amide bond (half-life >20 days). The extent of penetration for MTX–dextran conjugates was studied in three-dimensional human brain tumor cell cultures; penetration was significantly enhanced for MTX–dextran conjugates and the increased penetration was correlated with the stability of the linkage. These results suggest that modification of existing drugs may increase their efficacy against brain tumors when delivered directly to the brain interstitium.

9.7 Conclusion

Controlled release polymer implants are a useful new technology for delivering drugs directly to the brain interstitium. This approach is already in clinical use for treatment of tumors [33], and could soon impact treatment of other diseases. The mathematical models described in this chapter provide a rational framework for analyzing drug distribution after delivery. These models describe the behavior of chemotherapy compounds very well and allow prediction of the effect of changing properties of the implant or the drug. More complex models are needed to describe the behavior of macromolecules, which encounter multiple modes of elimination and metabolism and are subject to the effects of fluid flow. In addition, variations on this approach may be useful for analyzing drug delivery in other situations.

References

1. Lieb, W. and W. Stein, Biological membranes behave as non-porous polymeric sheets with respect to the diffusion of non-electrolytes. *Nature*, 1969. 224:240–249.
2. Stein, W.D., *The Movement of Molecules across Cell Membranes*. 1967, Academic Press: New York.

3. Simpkins, J., N. Bodor, and A. Enz, Direct evidence for brain-specific release of dopamine from a redox delivery system. *Journal of Pharmaceutical Sciences*, 1985. 74:1033–1036.

4. Gregoriadis, G., The carrier potential of liposomes in biology and medicine. *The New England Journal of Medicine*, 1976. 295:704–710.

5. Cotzias, C.G., M.H. Van Woert, and L.M. Schiffer, Aromatic amino acids and modification of parkinsonism. *The New England Journal of Medicine*, 1967. 276:374–379.

6. Triguero, D., J.B. Buciak, J. Yang, and W.M. Pardridge, Blood-brain barrier transport of cationized immunoglobulin G: Enhanced delivery compared to native protein. *Proceedings of the National Academy of Sciences USA*, 1989. 86:4761–4765.

7. Tokuda, H., Y. Takakura, and M. Hashida, Targeted delivery of polyanions to the brain. *Proceedings of the International Symposium on Control. Rel. Bioact. Mat.*, 1993. 20:270–271.

8. Friden, P., L. Walus, G. Musso, M. Taylor, B. Malfroy, and R. Starzyk, Anti-transferrin receptor antibody and antibody-drug conjugates cross the blood-brain barrier. *Proceedings of the National Academy of Sciences USA*, 1991. 88:4771–4775.

9. Friden, P.M., L.R. Walus, P. Watson, S.R. Doctrow, J.W. Kozarich, C. Backman, H. Bergman, B. Hoffer, F. Bloom, and A.-C. Granholm, Blood-brain barrier penetration and *in vivo* activity of an NGF conjugate. *Science*, 1993. 259:373–377.

10. Neuwelt, E., P. Barnett, I. Hellstrom, K. Hellstrom, P. Beaumier, C. McCormick, and R. Weigel, Delivery of melanoma-associated immunoglobulin monoclonal antibody and Fab fragments to normal brain utilizing osmotic blood-brain barrier disruption. *Cancer Research*, 1988. 48:4725–4729.

11. Blasberg, R., C. Patlak, and J. Fenstermacher, Intrathecal chemotherapy: Brain tissue profiles after ventriculocisternal perfusion. *The Journal of Pharmacology and Experimental Therapeutics*, 1975. 195:73–83.

12. Yan, Q., C. Matheson, J. Sun, M.J. Radeke, S.C. Feinstein, and J.A. Miller, Distribution of intracerebral ventricularly administered neurotrophins in rat brain and its correlation with Trk receptor expression. *Experimental Neurology*, 1994. 127:23–36.

13. Mahoney, M.J. and W.M. Saltzman, Controlled release of proteins to tissue transplants for the treatment of neurodegenerative disorders. *Journal of Pharmaceutical Sciences*, 1996. 85(12):1276–1281.

14. Morrison, P.F., D.W. Laske, H. Bobo, E.H. Oldfield, and R.L. Dedrick, High-flow microinfusion: Tissue penetration and pharmacodynamics. *American Journal of Physiology*, 1994. 266:R292–R305.

15. Haller, M.F. and W.M. Saltzman, Localized delivery of proteins in the brain: Can transport be customized? *Pharmaceutical Research*, 1998. 15:377–385.

16. Dang, W. and W.M. Saltzman, Controlled release of macromolecules from a biodegradable polyanhydride matrix. *Journal of Biomaterials Science, Polymer Edition*, 1994. 6(3):291–311.

17. Dang, W., *Engineering Drugs and Delivery Systems for Brain Tumor Therapy*. 1993, The Johns Hopkins University: Baltimore, MD.

18. Wyatt, T.L. and W.M. Saltzman, Protein delivery from non-degradable polymer matrices, in *Protein Delivery-Physical Systems*, L. Saunders and W. Hendren, Editors. 1997, Plenum Press: New York, NY. pp. 119–137.

19. Bird, R.B., W.E. Stewart, and E.N. Lightfoot, *Transport Phenomena*. 1960, New York: John Wiley & Sons. 780.

20. Patlak, C. and J. Fenstermacher, Measurements of dog blood-brain transfer constants by ventriculocisternal perfusion. *American Journal of Physiology*, 1975. 229:877–884.

21. Nicholson, C., Diffusion from an injected volume of a substance in brain tissue with arbitrary volume fraction and tortuosity. *Brain Research*, 1985. 333:325–329.

22. Morrison, P. and R.L. Dedrick, Transport of cisplatin in rat brain following microinfusion: An analysis. *Journal of Pharmaceutical Sciences*, 1986. 75:120–128.

23. Saltzman, W.M. and M.L. Radomsky, Drugs released from polymers: Diffusion and elimination in brain tissue. *Chemical Engineering Science*, 1991. 46:2429–2444.

24. Fung, L., M. Shin, B. Tyler, H. Brem, and W.M. Saltzman, Chemotherapeutic drugs released from polymers: Distribution of 1,3-bis(2-chloroethyl)-1-nitrosourea in the rat brain. *Pharmaceutical Research*, 1996. 13:671–682.

25. Dang, W. and W.M. Saltzman, Dextran retention in the rat brain following controlled release from a polymer. *Biotechnology Progress*, 1992. 8:527–532.

26. Krewson, C.E., M. Klarman, and W.M. Saltzman, Distribution of nerve growth factor following direct delivery to brain interstitium. *Brain Research*, 1995. 680:196–206.

27. Krewson, C.E. and W.M. Saltzman, Transport and elimination of recombinant human NGF during long-term delivery to the brain. *Brain Research*, 1996. 727:169–181.

28. Strasser, J.F., L.K. Fung, S. Eller, S.A. Grossman, and W.M. Saltzman, Distribution of 1,3-bis(2-chloroethyl)-1-nitrosourea (BCNU) and tracers in the rabbit brain following interstitial delivery by biodegradable polymer implants. *The Journal of Pharmacology and Experimental Therapeutics*, 1995. 275(3):1647–1655.

29. Fung, L.K., M.G. Ewend, A. Sills, E.P. Sipos, R. Thompson, M. Watts, O.M. Colvin, H. Brem, and W.M. Saltzman, Pharmacokinetics of interstitial delivery of carmustine, 4-hydroperoxycyclo-phosphamide, and paclitaxel from a biodegradable polymer implant in the monkey brain. *Cancer Research*, 1998. 58:672–684.

30. Jain, R.K., Barriers to drug delivery in solid tumors. *Scientific American*, 1994. 271(1): 58–65.

31. Clauss, M.A. and R.K. Jain, Interstitial transport of rabbit and sheep antibodies in normal and neoplastic tissues. *Cancer Research*, 1990. 50:3487–3492.

32. Dang, W.B., O.M. Colvin, H. Brem, and W.M. Saltzman, Covalent coupling of methotrexate to dextran enhances the penetration of cytotoxicity into a tissue-like matrix. *Cancer Research*, 1994. 54:1729–1735.

33. Brem, H., S. Piantadosi, P.C. Burger, M. Walker, R. Selker, N.A. Vick, K. Black et al. Placebo-controlled trial of safety and efficacy of intraoperative controlled delivery by biodegradable polymers of chemotherapy for recurrent gliomas. *Lancet*, 1995. 345:1008–1012.

10

Surfactant Transport and Fluid–Structure Interactions during Pulmonary Airway Reopening

David Martin
Tulane University

Anne-Marie Jacob
Tulane University

Donald P. Gaver III
Tulane University

10.1 Introduction 10-1
    Clinical Significance of Liquid Lining Flows
10.2 Fluid–Structure Interactions in the Lung 10-6
    Airway Closure • Airway Reopening • Biological Responses
    to Micromechanical Stress Field • Surfactant Physicochemical
    Interactions
Acknowledgment 10-13
References 10-13

10.1 Introduction

The primary function of the lung is to transport and exchange gas efficiently from the outside environment to the alveolar-capillary network. The highly complex and delicate network of bifurcating airways provides the conduit for this exchange. These airways consist of compliant tissue that is lined with epithelial cells that vary in structure and function in different regions of the lung. This network of airways consists of ~20 generations that bifurcate to create a vast membrane (approximately the size of a squash court) for gas exchange. The first 16 generations are considered the conducting airways, which primarily function to environmentally condition the air as it is delivered to the terminal regions of the lung. This conducting zone meets the respiratory zone at the smaller airways known as bronchioles and terminal bronchioles. Alveoli begin to arise along the airway after the 17th generation and as the number of alveoli progressively increases at each generation, the function of the airways begins to gradually shift from gas transport to gas exchange. The respiratory bronchioles diverge further into alveolar ducts and eventually terminate into alveolar sacs where whole clusters of alveoli are located. A total of ~300 million alveoli provide an expansive surface area (~70 m^2) and therefore serve as the primary site of gas exchange. The pulmonary capillary network that surrounds the alveoli is thinly separated by the alveolar-capillary barrier (100–200 nm thick), and thus is very fragile. A single layer of pulmonary capillary endothelial cells lines the capillary network on one side of the barrier and on the opposing side, pulmonary epithelial cells line the respiratory airways and alveoli. In a healthy lung, this barrier forms an impermeable membrane that is essential in maintaining separation between the airspace and the interstitial and blood fluids. Disruption of this barrier allows transport of excess fluid into the lung, leading to pulmonary edema that may result in respiratory failure. In addition to comprising part of the blood–gas barrier, the pulmonary epithelium is responsible for coating the walls of the airways and alveoli with a thin layer of serous fluid. As the largest epithelial surface in the body, a fluid barrier is necessary to protect the lung from the outside environment. However, more relevant to this discussion is the significant role that the lining fluid plays in the global mechanics of respiration.

It is now commonly understood that surface tension forces within the thin layer of lining fluid that coats the respiratory airways and alveoli are fundamental to the mechanical characteristics of the lung. To illustrate this, Figure 10.1 shows the pressure–volume relationship of one lung inflated and deflated with air and the other with saline (Neergaard 1929). The air-inflated lung demonstrates that for any volume; a substantially higher pressure is required during inflation as compared to deflation. In contrast, the inflation and deflation limbs of the saline-filled lung are nearly coincident. When a system behaves differently during the application than it does during the removal of a force, it is said to have a *hysteresis*. The lower pressures and small hysteresis area of the liquid-filled lung in comparison to the air-filled lung demonstrates that surface tension forces provide the lung with the majority of its elastic recoil.

Fundamentally, surface tension is of critical importance to both normal and pathophysiological pulmonary mechanics because it provides, at the microscale level, a cohesive property that functions to minimize the size of the air–liquid interface (Figure 10.2). As a result, surface tension serves to reduce the overall surface area of the liquid lining that coats the alveoli. To compensate, a static pressure difference across the air–liquid interface must be established to maintain the alveolar structure. This pressure difference follows the Young–Laplace relationship:

$$\Delta P = \frac{2\gamma}{R}, \tag{10.1}$$

where ΔP is the pressure difference across the air–liquid interface, γ is the surface tension, and R is the interfacial radius of curvature. At the micromechanical level, this pressure difference must be overcome to inflate the alveoli and so it is understood that surface tension opposes lung inflation. Von Neergaard's elegant experiments, by comparing the inflation pressures required to inflate the lung using either saline (removing the air–liquid interface) or air demonstrated the importance of the air–liquid interface in defining the macromechanical properties of the lung.

However, when looking at two interconnected alveoli of different radii (Figure 10.3), this relationship could present a potential instability to the pulmonary system. If surface tension was constant, the pressure would be higher in the smaller alveolus than a neighboring large alveolus. This interalveolar

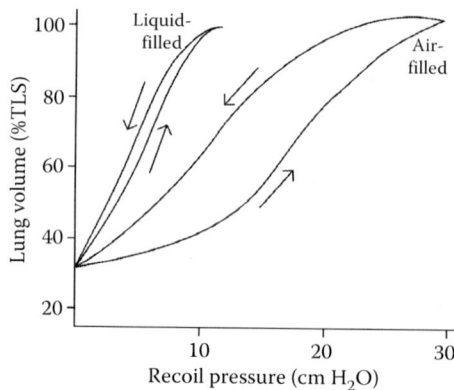

FIGURE 10.1 Pressure–volume hysteresis loops for air-filled and liquid-filled lungs.

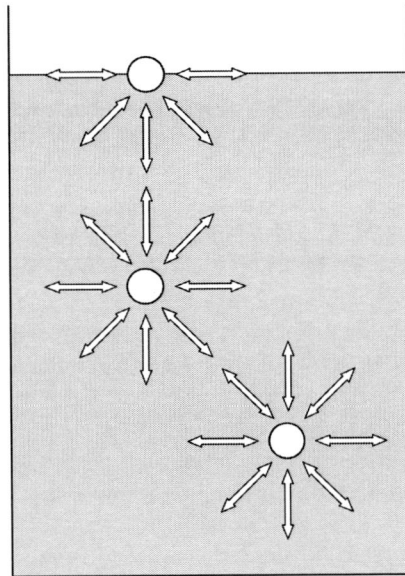

FIGURE 10.2 The cohesive forces between liquid molecules are balanced within the bulk phase but remain unopposed at the surface. The net result is an inward (bulk-directed) force on surface molecules and surface tension along the air–liquid interface.

Alveolar instability and the counteractive effects of pulmonary surfactant ℬ

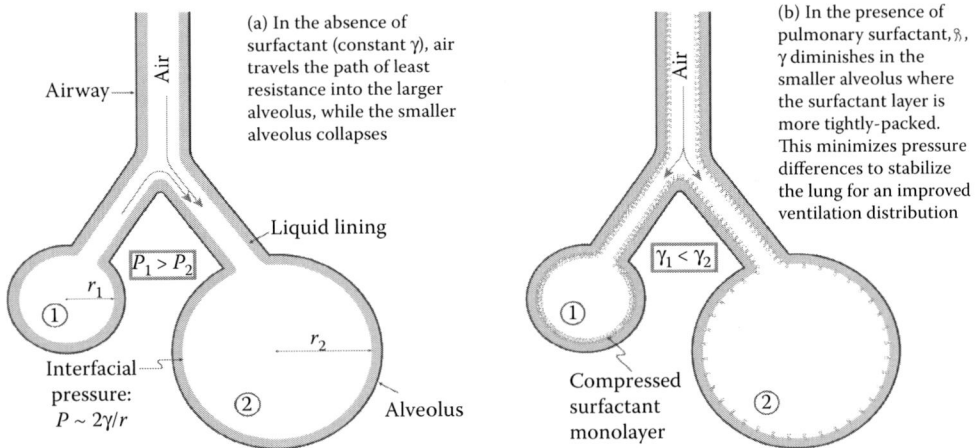

(a) In the absence of surfactant (constant γ), air travels the path of least resistance into the larger alveolus, while the smaller alveolus collapses

Airway

Air

Liquid lining

$P_1 > P_2$

r_1

①

r_2

Interfacial pressure: $P \sim 2\gamma/r$

②

Alveolus

(b) In the presence of pulmonary surfactant, ℬ, γ diminishes in the smaller alveolus where the surfactant layer is more tightly-packed. This minimizes pressure differences to stabilize the lung for an improved ventilation distribution

Air

$\gamma_1 < \gamma_2$

①

Compressed surfactant monolayer

②

FIGURE 10.3 Pulmonary surfactant plays a vital stabilizing role in pulmonary mechanics. (a) In the absence of pulmonary surfactant, airflow is directed into the larger alveolus where pressures are lowest ($P_2 < P_1$), resulting in collapse and overdistension of the smaller and larger alveoli, respectively. (b) In the healthy lung with functional pulmonary surfactant, the increased interfacial density of pulmonary surfactant molecules in the compressed smaller alveolus reduces the surface tension. This reduces the insufflation pressure and stabilizes the lung.

pressure gradient would drive the flow from the smaller alveolus to the larger alveolus and could consequently lead to an avalanche effect that would eventually cause the collapse of nearly all interconnected alveoli. However, as described below, under normal conditions, this mechanical instability is prevented by the presence of surfactant in the lining fluid of the lungs that dynamically alters the surface tension of the air–liquid interface and stabilizes the lung.

Surfactant is a lipid–protein complex that is synthesized and released by alveolar type II epithelial cells. This complex surface-active compound contains both hydrophobic and hydrophilic regions to allow the molecule to spontaneously adsorb to and form monolayers along the air–liquid interface. The role of surfactant in pulmonary fluid mechanics depends on its natural ability to disrupt intermolecular forces by interfering with the attractive forces between water molecules at the interfacial surface—thus lowering the surface tension. While this surfactant mixture is largely comprised of lipids (90%), the surfactant proteins (10%) are required for normal function (Hall et al. 1992; Yu and Possmayer 1993). Finally, the molecule dipalmitoyl phosphatidylcholine (DPPC) makes up 80% of the phospholipid and is largely responsible for the ultra-low surface tensions necessary for respiratory function (<5 dyn/cm) (Klaus et al. 1961; Hawco et al. 1981; Tchoreloff et al. 1991).

The potential mechanical instability described above (Figure 10.3) demonstrates the need for surface tension to be dynamically lowered in regions of low surface area and increased in regions of high surface area to maintain the stability of the lung. This is, in fact, exactly how surfactant works. The molecular interactions at the air–liquid interface are dependent on the surface concentration at the interface. The interfacial concentration is, in turn, dependent on the surface area available, which naturally changes during inspiration and expiration. During expiration, the rapid decrease in surface area causes compression of surfactant at the interface until the surface saturates and causes surfactants adsorbed to the air–liquid interface to collapse into the bulk fluid where they form subsurface multilayers (Schurch et al. 1998; Bachofen et al. 2005). This process is shown in Figure 10.4. Through this process, the interfacial concentration is at its highest when alveoli are small, which substantially reduces the surface tension and helps to equilibrate the pressure difference between large and small alveoli.

Critical to surfactant function is the ability of surfactant located in the lining fluid to distribute itself between the interface and liquid phases—this process is very important to the overall mechanical function of the lung. For example, as the area of the air–liquid interface decreases during exhalation, surfactant is exchanged from the interface to a subsurface multilayer that serves as a reservoir. During inspiration, the increasing surface area rarefies the interfacial concentration and thus allows for surfactant to be reincorporated along the interface. This reincorporation is critically dependent on adequate surfactant protein function (Hall et al. 1992). By changing the surface tension during inspiration and expiration, surfactant not only has a stabilizing effect on the small respiratory airways and alveoli, but also significantly decreases the inflation pressure necessary for inspiration. Understanding the coupling of surfactant transport and fluid mechanical properties may be critical to elucidating disease mechanisms and the development of treatment and prevention strategies.

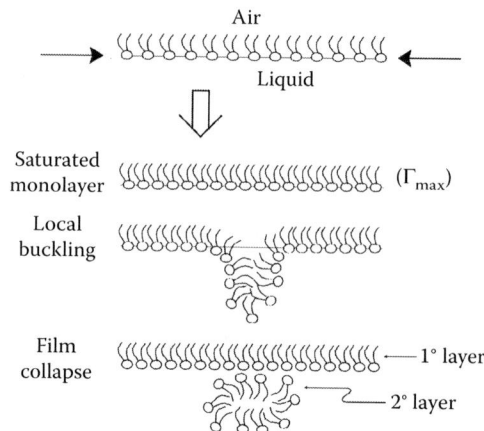

FIGURE 10.4 A surfactant monolayer compressed beyond saturation (Γ_{max}) will undergo "squeeze out," wherein a portion will buckle into the liquid and break off to form a secondary layer.

10.1.1 Clinical Significance of Liquid Lining Flows

It has been discussed that for healthy respiratory function, the lungs must constantly maintain an impermeable membrane between the airways and interstitial fluids, and they must have a properly functioning surfactant system capable of keeping surface tensions in the lung low enough to facilitate breathing. However, severe respiratory illnesses caused by either an initiating insult to the lung or immature development of the lung threaten both the blood–gas barrier and functioning levels of surfactant. These respiratory illnesses include infant respiratory distress syndrome (IRDS), acute or adult respiratory distress syndrome (ARDS), and acute lung injury (ALI). Conditions resulting from these illnesses often lead to disruption of the alveolar-capillary membrane, insufficient or deactivated surfactant, and subsequent airway and alveolar closure (atelectasis). Collectively, this can lead to significant damage to sensitive tissues of the lung with potentially grave consequences.

10.1.1.1 IRDS

Infant respiratory distress syndrome (IRDS) is a life-threatening condition that afflicts premature infants born prior to the seventh month of gestation. Respiratory distress in these patients is due to insufficient or abnormal surfactant production, which is the result of a structurally and functionally immature lung (Avery and Mead 1959). Immediately following birth, the distal regions of the lung including the respiratory airways and alveoli are filled with amniotic fluid. To begin breathing properly, the infant must generate sufficient inflation pressures to force air into the lungs, clear the fluid-filled airways, and ventilate the most distal regions of the lung. A healthy infant with fully functioning surfactant systems is easily capable of opening the lung and preventing subsequent airway collapse during expiration. However, infants born before 25 weeks of gestation are likely to have lower levels of surfactant (Usher et al. 1971). Elevated surface tensions due to diminished interfacial surface activity is compounded by the tiny radius of curvature imposed by the underdeveloped lung, resulting in enormous inflation pressures (Gaver et al. 1990, 1996; Halpern and Grotberg 1993). Consequently, premature neonates are generally unable to autonomously generate high-enough inflation pressures to completely clear the lung of fluid. In addition, without sufficient surfactant, regions of the lung that are opened are unstable and therefore prone to collapse (Levitzky 1982). This can lead to inhomogeneous ventilation, atelectic regions of the lung, and eventually, airway closure, hypoxia, and death (Finlay-Jones et al. 1974). Respiratory distress syndrome presents within the first few minutes after birth and requires immediate intensive care treatment to prolong the life of the child. This generally includes a combination of supportive therapies involving provision of exogenous surfactant (surfactant replacement therapy), high-frequency ventilation, and/or positive end-expiratory pressure ventilation. Despite a 60% decrease in infant mortality rates since 1989 due to advancements in this field, IRDS remains the most common cause of respiratory failure and is the seventh leading cause of death in premature infants today (Guyer et al. 1997; Heron et al. 2009).

10.1.1.2 ALI–ARDS

ALI and the more severe ARDS are the result of direct or indirect injury to the pulmonary epithelium and endothelium, and are characterized by edema, atelectasis, inflammation, and end-stage fibrosis (Bernard et al. 1994). ALI and ARDS are commonly caused by both direct and indirect affronts, including primary pneumonia, aspiration of gastric contents, sepsis from a nonpulmonary source, and multiple trauma (Ware and Matthay 2000). Respiratory failure is initiated by injury to the alveolar-capillary membrane. The alveolar epithelium is composed of two predominant cell types: type I and type II alveolar epithelial cells (EpC). The type I cells have a flattened morphology and account for 95% of the alveolar surface area. Therefore, they are primarily responsible for maintaining the impermeable blood–gas barrier. The type II cells have a cuboidal morphology and account for the remaining 5% of the alveolar surface area. They are responsible for fluid transport, reepithelialization, and most importantly to this topic, production of surfactant (Ross and Pawlina 2006). Hence, destruction of the alveolar epithelium has compounding consequences. Following injury to the lung, type II cells can die or differentiate into type I cells. This decreases the overall production of surfactant in the lung as well as interfering with normal transport of edema fluid out of the lung. Simultaneously, type I cell death and injury causes severe disruption of the alveolar-capillary membrane, allowing proteinaceous edema fluid to flood the alveolar airspace. Proteins, neutrophils, and macrophages present in the alveoli interact with and deactivate the already diminished surfactant in the fluid lining (Ware and Matthay 2000). In addition, when the epithelium is unable to sufficiently repair itself, formation of hyaline membranes occurs along denuded airways, decreasing the compliancy of the lung (Kim and Malik 2003). With increasing pulmonary edema, collapsed airways and alveoli, stiffening of the lungs, and high interfacial surface tensions, mechanical ventilation becomes necessary to maintain adequate gas exchange. However, treatment for ALI and ARDS is still in vast need for improvement as reflected by recently found mortality rates of ~39% and 42%, respectively, with an annual number of adult deaths analogous to that associated with breast cancer or AIDS (Rubenfeld et al. 2005). It is estimated that there are 190,600 cases per year, leading to 74,500 deaths and 3.6 million hospital days (2.2 million days spent in ICU) in the United States (Rubenfeld et al. 2005).

10.1.1.3 VILI

Despite recent advancements in alternative modalities such as extracorporeal membrane oxygenation and surfactant replacement therapy, mechanical ventilation remains the standard treatment of care for cases of ARDS and IRDS. Although ventilation is successful in recruiting collapsed airways and restoring blood oxygenation, the mechanical stresses associated with mechanical ventilation have been shown to exacerbate lung injury (Dreyfuss and Saumon 1998). Ventilator-induced lung injury (VILI) is a clinical condition characterized by the collective damaging processes caused by high lung volume ventilation (volutrauma), the repeated closure and reopening of fluid-filled structures of the lung (atelectrauma), and mechanically induced systemic inflammatory responses (biotrauma). During high tidal volume ventilation, overdistension of the alveolar wall subjects the basement membrane and pulmonary epithelium to large stretching deformations that can lead to cellular injury, stress failure of the plasma membrane, and apoptosis (Cavanaugh and Margulies 2002; Gajic et al. 2003; Syrkina et al. 2008). During atelectrauma, cyclic closure and reopening of fluid occluded airways introduces interfacial flows that subject the pulmonary epithelium to large mechanical stresses as the airways are reopened and cleared of fluid. These damaging processes further the injury already done to the lung by increasing the permeability of the epithelium, worsening pulmonary edema, decreasing surfactant production, and deactivating or washing away any remaining surfactant (Robertson 1984). While mechanical ventilation is necessary for patient treatment, the associated mechanical insults may exacerbate damage to the lung. In addition, it is now recognized that this may contribute to large-scale inflammatory responses that can contribute to multiorgan failure (Montgomery et al. 1985; Ferring and Vincent 1997).

10.2 Fluid–Structure Interactions in the Lung

Healthy, normal function of the lung depends on the interaction between the delicate pulmonary tissue, the lining fluid that coats the airway walls, and pulmonary surfactant that dynamically modulates surface tension within the lung. In instances where acute respiratory illnesses have altered the properties of the lung, these fluid–structure interactions become critically important to the patient's pathological condition. These damaging forces are most commonly inflicted during the repetitive recruitment and derecruitment of pulmonary airways during mechanical ventilation of the lung as described below.

10.2.1 Airway Closure

Respiratory function is highly dependent on fast, efficient transport of gas from the mouth to the alveoli. Free passage of gas through the airways is maintained by mechanical properties of the airway structures and low interfacial surface tension. Airway closure generally occurs due to structural collapse, or liquid bridges created by surface-tension-induced instabilities that can obstruct the airway (Macklem et al. 1970). This can become serious when airway closure affects significant portions of the lung and interferes with critical gas exchange. In healthy human lungs, airway closure of very small peripheral airways may occur during end-exhalation (Macklem et al. 1970); however, these occlusions are typically cleared during inspiration. More substantial airway closure for longer periods of time usually occurs in pathological conditions such as ARDS and IRDS due to higher surface tensions and pulmonary edema. Airway closure is fundamentally a fluid blockage or liquid bridge that prevents passage of air. There are generally two mechanisms by which airway closure can occur, "compliant collapse" and "film collapse" (Heil et al. 2008). In a purely fluid mechanical "film collapse," the compressive force on the airway wall due to surface tension forces is not significant compared to the airway stiffness. Thus, no considerable deformations are possible. However, if the film thickness is great enough, a surface tension-induced instability (Raleigh-Plateau) can cause the spontaneous redistribution of the thin fluid lining into a liquid bridge (Kamm and Schroter 1989). In "compliant collapse," increased compliance of the airway walls coupled with high surface tensions can cause significant airway deformations. In this scenario, the tendency of surface tension to minimize the surface area of the air–liquid interface operates as a buckling force. This can result in either axisymmetric or nonaxisymmetric collapse in which the collapsed airway is then able to form a fluid occluding liquid bridge with relatively small volumes of liquid (Halpern and Grotberg 1992, 1993; Hill et al. 1997; Heil 1999; Rosenzweig and Jensen 2002). This is more common in small airways and alveoli since the airway compliance increases with each generation (Hughes et al. 1970).

10.2.2 Airway Reopening

The fluid–structure interactions responsible for lung injury during airway reopening are fundamentally multiscale. At the macroscale (organ level), large upstream pressures required to recruit collapsed airways can result in the overdistension and damage to patent regions of lung that were previously unaffected. The "yield" pressure ($P_{yield} \sim 8\gamma/R$) represents the upstream pressure required to inflate collapsed airways (Gaver et al. 1990). In order to investigate the effect of respiratory illnesses such as ARDS on global lung function, the yield pressure for obstructed airways was estimated in adults with healthy surfactant function and compared to the yield pressures required for adults with a surfactant deficiency. The results showed a healthy yield pressure of $P_{yield} \sim 5$ cm H_2O and a disease state yield pressure of $P_{yield} \sim 15$–20 cm H_2O. Also by comparison, premature infants with surfactant deficiencies were estimated to have $P_{yield} \sim 50$ cm H_2O. These pressures are applied without discrimination to the whole lung even though airway closure is a local event. Therefore, recruitment of collapsed airways can be extremely harmful to patent regions of the lung because it may lead to large stretching deformations due to hyperinflation.

On the microscale (tissue level), surface tension forces located at the air–liquid interface of the progressing bubble subject the EpC lining of closed airways and alveoli to a nonuniform, time-dependent, micromechanical stress field. In order to predict the magnitude of the microscale stresses imparted by the interfacial flows, fluid mechanical models are developed that are founded upon fundamental investigations of multiphase flows.

During ventilation of a fluid-occluded airway, a progressing front of air is propagated through the closed airway. In compliant models the bubble "peels apart" any collapsed walls and displaces the surrounding fluid. In rigid and compliant reopening models, this introduces a semi-infinite bubble into the lung airways as illustrated in Figure 10.5. These models are based upon classical studies of interfacial flows in the hydrodynamic literature related to bubble propagation in a rigid capillary tube (Bretherton 1961; Taylor 1961). The mechanics of reopening a liquid-occluded airway, where inertial effects are negligible, are governed by the Stokes and continuity equations (Reinelt and Saffman 1985).

$$\nabla p = Ca \nabla^2 \mathbf{u}, \quad \nabla \cdot \mathbf{u} = 0, \tag{10.2}$$

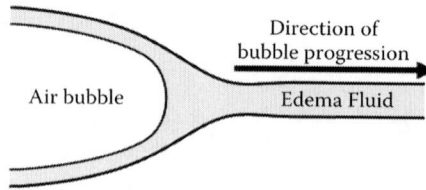

FIGURE 10.5 During airway reopening of collapsed and fluid-occluded regions of the lung, a semi-infinite bubble is forced from left to right as it peels apart the airway walls and displaces the surrounding fluid.

where **u** and p are dimensionless fluid velocity and pressure and Ca is the dimensionless representation for fluid velocity, which relates viscous to surface-tension forces, called the capillary number:

$$Ca = \mu U/\gamma, \tag{10.3}$$

where μ and γ are viscosity and surface tension and U is the bubble tip velocity. Ca is a fundamental dimensionless parameter in the flow Equation 10.2, indicating that airway reopening is highly dependent on the competing effects of viscosity and surface tension.

In models of reopening that are based upon rigid airway walls and a constant surface tension interface, the stress and kinematic boundary conditions along the air–liquid interface are

$$[\,\sigma \cdot \hat{\mathbf{n}}\,] = (\gamma\kappa)\hat{\mathbf{n}}, \quad \mathbf{u} \cdot \hat{\mathbf{n}} = 0. \tag{10.4}$$

Here, [ ] represents a jump condition at the interface, $\hat{\mathbf{n}}$ is the outward-facing normal vector, κ is the interfacial curvature, and $\sigma = -p\mathbf{I} + \nabla\mathbf{u} + \nabla\mathbf{u}^T$ is the stress tensor.

At the airway wall, the boundary condition is either a no-slip condition (for rigid models) or

$$[\,\hat{\mathbf{m}} \cdot \sigma \cdot \hat{\mathbf{m}}\,] = -\eta\kappa + \phi(y + 1)m_y, \tag{10.5}$$

where $\hat{\mathbf{m}}$ is the unit outward-facing normal vector for the wall, $\eta = T/\gamma$ is the dimensionless wall tension, and $\phi = KH^2/\gamma$ is the dimensionless wall elasticity. This coupled fluid–structure system alters the mechanical environment near the bubble tip in a spatial and time-dependent manner that, in a constant surface tension system, increases the damaging mechanical stresses on the lung epithelium.

Figure 10.6a depicts the stress field that occurs during the reopening of a compliantly collapsed airway. In this figure, a semi-infinite bubble passes from left to right, peeling apart the opposing walls that are held in apposition by a viscous fluid (Gaver et al. 1996). Figure 10.6b shows the stress field exerted on the EpC lining as a semi-infinite bubble clears a fluid-occluded, rigid airway. In both models, the EpC lining is exposed to rapid, time-dependent changes in normal and shear stresses as the bubble progresses downstream. The compliantly collapsed airway model has slightly more complex behavior near the bubble tip due to added pressures caused by the deformation of the basement membrane. As the bubble approaches the cells, they experience an inward normal and backward-facing shear stress. As the bubble passes directly over the cells, the forces are reversed so that the cells instead experience an outward normal and forward-facing shear stress. Finally, cells far upstream (the region covered by the bubble) experience a constant outward normal stress while cells far downstream experience no significant stress at all.

In a rigid, parallel-plate model of airway reopening, cells experience similar spatial and temporal gradients in shear and normal stresses but are not exposed to the inward normal and shear stresses located just downstream of the bubble tip. The spatial and temporal gradients of both shear and normal stresses are the most important characteristics of the system because the rapid changes in magnitude and

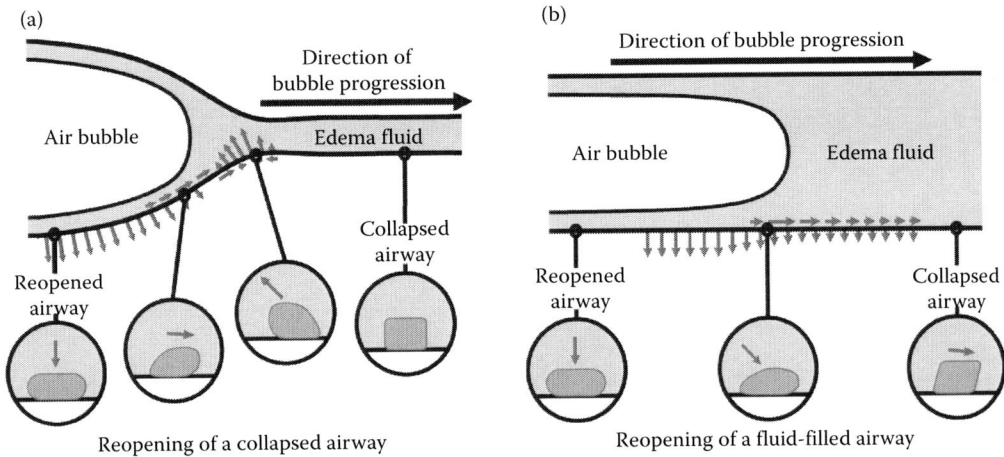

FIGURE 10.6 The reopening of atelectatic airways, either (a) collapsed or (b) fluid-filled, generates excessive fluid mechanical stresses and stress gradients that dramatically alter both the structure and function of the lining cellular epithelium. (Adapted from Bilek, A. M. et al. 2003. *J Appl Physiol* **94**: 770–783.)

direction pose the greatest potential for damage to the lung epithelium. In rigid surfactant-free systems, the magnitudes of the tangential- and normal-stress field have been predicted to be (Kay et al. 2004):

Shear stress (τ_s)

$$(\tau_s)_{max} = 0.69 \frac{(\mu U)^{0.36} \gamma^{0.64}}{H}. \qquad (10.6)$$

Shear stress gradient ($d\tau_s/dx$)

$$(d\tau_s/dx)_{max} = \frac{0.22\gamma}{H^2} + 1.2 \frac{(\mu U)^{0.75} \gamma^{0.25}}{H^2}. \qquad (10.7)$$

Normal-stress gradient (dP/dx)

$$(dP/dx)_{max} = 0.34 \frac{\gamma^{1.29}}{(\mu U)^{0.29} H^2}. \qquad (10.8)$$

Stress exposure duration ($\Delta t_{exposure}$)

$$\Delta t_{exposure} = 2.94 \frac{H\mu^{0.29}}{U^{0.71} \gamma^{0.29}}. \qquad (10.9)$$

10.2.3 Biological Responses to Micromechanical Stress Field

Understanding the biological responses of lung tissue to the micromechanical stress field near the bubble tip during airway reopening is fundamental to the issue of atelectrauma-induced VILI. As described above and illustrated in Figure 10.6, the stress field can be decomposed into four potentially injurious components—shear stress, the shear stress gradient, pressure, and the pressure gradient. The potential

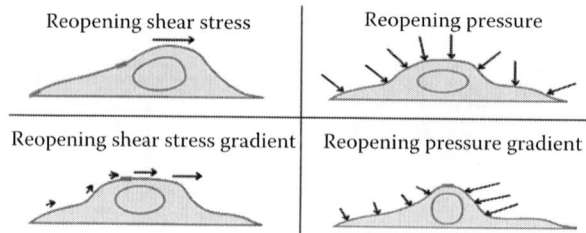

FIGURE 10.7 Airway reopening subjects pulmonary epithelial cells to a complex, time-dependent micro-mechanical stress field that can be decomposed into four separate injurious stress components.

damaging mechanisms of these stress components are illustrated in Figure 10.7 and briefly described below. Interestingly, experimental results have shown that slower bubble velocities result in higher levels of cell death and therefore, in determining the primary mechanism of damage, it should follow that the magnitude of the dominating stress should increase with decreasing velocities (Bilek et al. 2003). To this end, Bilek et al. evaluated each of the four primary stresses to see if an increase in their magnitude correlated directly to an increase in cellular death, thus indicating that stress component as the primary mechanism of epithelial cell damage.

- Shear stress—The maximum shear stress occurs in the region just upstream of the tip, causing fluid to flow over the surface of the cell and apply a tractional force that could induce cell deformations capable of causing plasma membrane disruptions. In addition, it is possible that the cell membrane may become "rarified" as shear stresses cause nontethered components of the membrane to shift. This may have the effect of causing the cell membrane to be easily torn (Vlahakis and Hubmayr 2000). Under these assumptions, the magnitude of damage should correlate to faster velocities. Since experimental observations have shown that slower velocities correlate to a larger extent of cell damage, shear stress can largely be discounted as the primary stress component responsible for cellular damage.
- Shear stress gradient—Due to the dynamic nature of the stress field exerted on a single cell as the bubble passes over its surface, cells are exposed to a gradient of shear stress. This nonuniform stress field causes force imbalances within the plane of the cell membrane. Cells have a small tolerance to tension within the cell membrane and therefore may easily rupture (Vlahakis and Hubmayr 2000). Although larger stress gradients could explain the increased damage seen in surfactant-deficient airway reopening models, they do not show any difference in damage at lower velocities and therefore cannot qualify as the dominating mechanism of damage.
- Pressure—The maximum pressure occurs in the patent regions upstream of the bubble tip. The pressure is an evenly distributed normal stress directed outward on the cells. While this pressure may be relatively large, it is unlikely to seriously damage the cells as the transmural pressure will quickly equilibrate to the applied pressure. In addition, the maximum reopening pressure increases with velocity and therefore does not correlate with the observed increased damage at low velocities.
- Pressure gradient—Similar to the shear stress gradient, a single cell is exposed to changing magnitudes of normal stresses across the length of the cell. It is hypothesized that the flat regions on either side of the protruding nucleus should experience a flattening of the cell and stretching of the membrane. Within the cell, it is speculated that the pressure gradient may induce internal flows that could rupture the membrane from within. Finally, the high-profile region of the protruding nucleus is subjected to net normal forces acting on either side, resulting in a "pinching" of that region. This could tear the membrane at either the base of the protrusion or rupture the top surface of the cell. Experimental results show that slower bubble velocities create larger pressure gradients across the cell body while faster bubble velocities decrease the pressure gradient across the cell body.

In summary, an increase in pressure difference correlates with the increase in damage seen at lower velocities. Furthermore, when tested in an airway reopening model with surfactant in the occlusion fluid, pressure gradients were reduced in correlation with a reduction in damage, further supporting pressure gradient as the dominating stress component responsible for cellular damage (Bilek et al. 2003). A follow-up study further substantiated this finding by discounting stress exposure duration as a mechanism of damage (Kay et al. 2004). Furthermore, additional studies reinforced pressure gradient as the determining factor for cell damage by targeting parameters such as channel height, subconfluent culture conditions, and repeated reopening events (Yalcin et al. 2007). These studies demonstrated that cell topography may play a large role in damage magnitude and distribution, which is consistent with the theoretical investigations of Jacob and Gaver (2005).

10.2.4 Surfactant Physicochemical Interactions

While fluid–structure interactions clearly create damaging mechanical stresses, surfactant physico-chemical interactions are critical to the protection of the airway epithelium. In order to investigate this important interaction during airway reopening, it is necessary to expand the models above by introducing the biophysical properties of surfactant located in the lining fluid. This includes surfactant transport in the bulk phase by convection and diffusion, and the sorption kinetics between the bulk phase and the air–liquid interface. The flow fields in the bulk fluid and sorption kinetics at the interface both contribute to the rate and manner in which surfactant adsorbs to the interface where it locally decreases surface tension. These interactions are depicted in Figure 10.8, and defined below by Equations 10.10 and 10.11.

Surfactant in the bulk phase (denoted by C) is transported in the liquid occlusion following the convection–diffusion equation:

$$\frac{\partial}{\partial x}(uC) + \frac{\partial}{\partial y}(vC) = \frac{1}{\text{Pe}_b}\left(\frac{\partial^2 C}{\partial x^2} + \frac{\partial^2 C}{\partial y^2}\right). \tag{10.10}$$

Here, Pe_b is the bulk Peclet number that represents the ratio of convection and diffusion of surfactant molecules. The convection field provides converging (diverging) stagnation points that accumulate (deplete) surfactant at the subsurface (Cs) and interface (Γ), as shown in Figure 10.8a.

The transport of surfactant molecules to the bubble surface is defined as

$$\frac{d}{ds}(u_s\Gamma) + \kappa_i u_n\Gamma = \frac{1}{\text{Pe}_s}\frac{d^2\Gamma}{ds^2} + j, \tag{10.11}$$

where u_n and u_s are the normal and tangential components of velocity along the interface, j is the flux of surfactant from the bulk to the interface, and Pe_s is the surface Peclet number that represents surfactant transport due to interfacial convection and diffusion. Models for j are given by Otis et al. (1994) and Krueger and Gaver (2000)—these models provide concentration-dependent fluxes that depend on the local subsurface and interfacial concentrations, and may incorporate surfactant multilayer interactions (Figure 10.4).

The coupling between fluid mechanics and transport processes described above introduces a nonuniform distribution of surfactant at the air–liquid interface (Figure 10.8b, left panel). The surface tension at the interface is directly related to the interfacial surfactant concentration, Γ, through an equation of state, that in linear form is provided by

$$\gamma = 1 - E_l(\Gamma - 1), \tag{10.12}$$

where E_l is the surface elasticity number. Therefore, a variation in Γ leads to the formation of surface tension variation as given by Figure 10.8b (right panel).

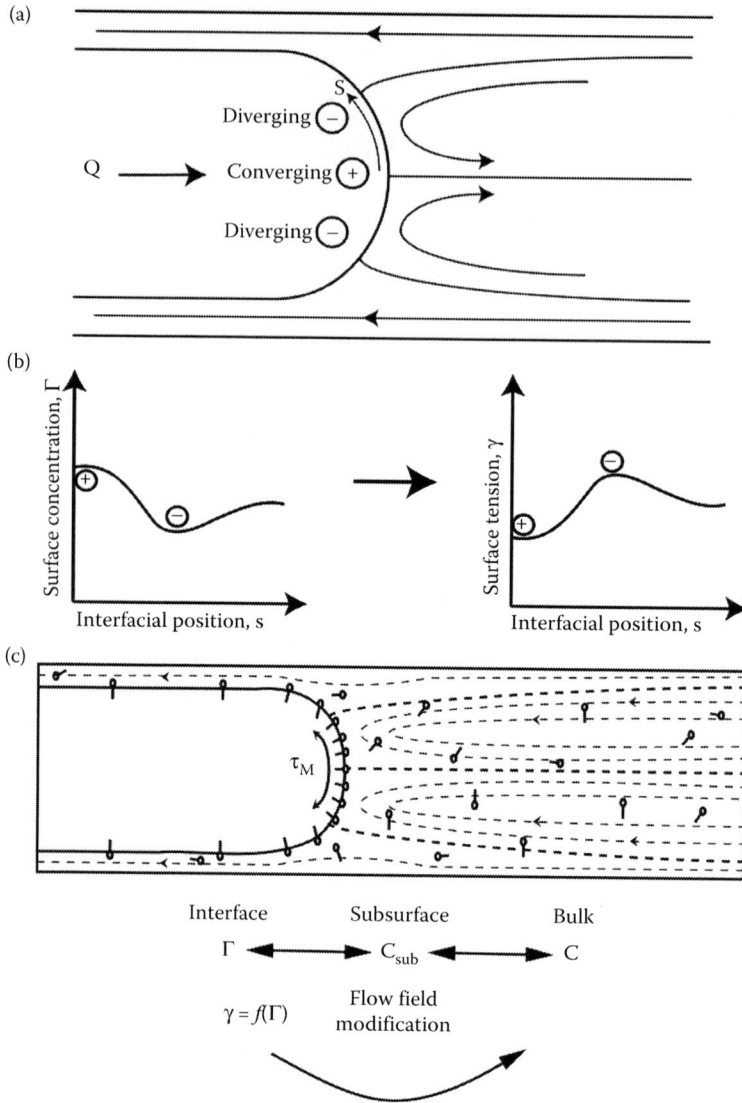

FIGURE 10.8 (a) A flow field surrounding a semi-infinite bubble as it propagates through a fluid-filled tube has converging (⊢) and diverging (–) stagnation points where surfactant accumulates and undergoes depletion, respectively. (b) The relationship between Γ (interfacial surfactant concentration), γ (interfacial surface tension), and the interfacial position coordinate s. Variations in Γ cause surface tension gradients along the interface. (c) Streamlines surrounding semi-infinite bubble as it propagates through a fluid-filled tube. Surfactant molecules located in the bulk fluid are transported via convection along these streamlines to the air–liquid interface where sorption kinetics cause nonuniform distribution of surfactant along the surface, resulting in Marangoni stresses. (From Pillert, J. E. and D. P. Gaver 2009. *Biophys J* **96**(1): 312–327.)

Variation in the surface tension introduces nonuniform mechanical stresses at the air–liquid interface, following the general form of the stress boundary condition at the air–liquid interface:

$$[\,\sigma \cdot \hat{\mathbf{n}}\,] = (\gamma\kappa)\hat{\mathbf{n}} + \frac{\partial\gamma}{\partial s}\,\hat{\mathbf{t}}\,. \tag{10.13}$$

Here, $\hat{t}$ is the unit tangential vector, s is the arc length vector, and $\partial\gamma/\partial s$ describes a tangential stress (termed "Marangoni stress").

Figure 10.8c combines the concepts of Figure 10.8a, b, and shows a closed loop region where surfactant is recirculated near the bubble tip. This results in higher surface concentrations at the bubble tip and nonuniform interfacial concentrations along the surface of the bubble. Variations in the dynamic surface tension along an interface exert the Marangoni stress (τ_M) as well as altering the pressure drop across the air–liquid interface as determined by the Young–Laplace law (Equation 10.1). Modification of the local surface tension at the interface creates a nonequilibrium normal stress that can alter the pressure required to push the bubble through the airway. Additionally, Marangoni stresses act tangential to the interface and are directed toward the region of higher surface tension. This results in a "rigidification" of the interface that in general retards the motion of the bubble.

To summarize, nonuniform adsorption of surfactant to the air–liquid interface causes surface tension along the interface to be modified in a spatial- and time-dependent manner. This, in turn, modifies the mechanical stress field in a manner that may either protect or damage pulmonary epithelium.

Experimental investigations of surfactant transport in rigid tube models of steady airway reopening indicate that transport limitations exist that can significantly increase reopening pressures (Ghadiali and Gaver 2000). However, by integrating the above physicochemical behavior of surfactant into the existing airway reopening models, insights into ideal ventilation strategies are possible, which may exploit the protective effects of surfactant and surfactant transport. These findings may be critical to the goal of diminishing the damaging effects of atelectrauma-induced VILI and thus improving the prognosis for patients suffering from IRDS or ARDS. For example, an early study by Ghadiali and Gaver (2003) simulated interfacial surfactant dynamics, which demonstrated the important physicochemical interactions that influence surfactant uptake and mechanical stresses in steady flow. Subsequent computational investigations by Zimmer et al. (2005) and Smith and Gaver (2008) demonstrate that pulsatility can be used to more efficiently deliver surfactant to the bubble tip region. Experiments by Pillert and Gaver (2009) were able to show that by superimposing a fast-forward pulsatile flow onto a fast steady flow, reopening pressures are substantially decreased due to increased surfactant transport and adsorption. This result demonstrates the promise of developing novel modes of mechanical ventilation in order to reduce the incidence or severity of ventilator-induced lung injury.

Acknowledgment

This work was supported by NIH grant R01-HL81266 and NSF grant CBET-1033619 (only NIH was in before).

References

Avery, M. E. and J. Mead 1959. Surface properties in relation to atelectasis and hyaline membrane disease. *AMA J Dis Child* **97**: 517–523.

Bachofen, H., U. Gerber et al. 2005. Structures of pulmonary surfactant films adsorbed to an air-liquid interface in vitro. *Biochim Biophys Acta Biomembranes* **1720**(1–2): 59–72.

Bernard, G. R., A. Artigas et al. 1994. The American-European consensus conference on ards—definitions, mechanisms, relevant outcomes, and clinical-trial coordination. *Am J Respir Crit Care Med* **149**(3): 818–824.

Bilek, A. M., K. C. Dee et al. 2003. Mechanisms of surface-tension-induced epithelial cell damage in a model of pulmonary airway reopening. *J Appl Physiol* **94**: 770–783.

Bretherton, F. P. 1961. The motion of long bubbles in tubes. *J Fluid Mech* **10**(02): 166–188.

Cavanaugh, K. J. and S. S. Margulies 2002. Measurement of stretch-induced loss of alveolar epithelial barrier integrity with a novel *in vitro* method. *Am J Physiol Cell Physiol* **283**: 1801–1808.

Dreyfuss, D. and G. Saumon 1998. Ventilator-induced lung injury: Lessons from experimental studies. *Am J Respir Crit Care Med* **157**: 294–323.

Ferring, M. and J. Vincent 1997. Is outcome from ARDS related to the severity of respiratory failure? *Eur Respir J* **10**(6): 1297–1300.

Finlay-Jones, J., J. M. Papadimitriou et al. 1974. Pulmonary hyaline membrane: Light and electron microscopic study of the early stage. *J Pathol* **112**: 117–124.

Gajic, O., J. Lee et al. 2003. Ventilator-induced cell wounding and repair in the intact lung. *Am J Respir Crit Care Med* **167**: 1057–1063.

Gaver, D. P., R. W. Samsel et al. 1990. Effects of surface tension and viscosity on airway reopening *J Appl Physiol* **69**: 74–85.

Gaver, D. P., R. W. Samsel et al. 1996. The steady motion of a semi-infinite bubble through a flexible-walled channel. *J. Fluid Mech* **319**: 25–65.

Ghadiali, S. N. and D. P. Gaver 2000. An investigation of pulmonary surfactant physicochemical behavior under airway reopening conditions. *J Appl Physiol* **88**(2): 493–506.

Ghadiali, S. N. and D. P. Gaver 2003. The influence of non-equilibrium surfactant dynamics on the flow of a semi-infinite bubble in a rigid cylindrical capillary tube. *J Fluid Mech* **478**: 165–196.

Guyer, B., J. A. Martin et al. 1997. Annual summary of vital statistics—1996. *Pediatrics* **100**(6): 905–918.

Hall, S. B., A. R. Venkitaraman et al. 1992. Importance of hydrophobic apoproteins as constituents of clinical exogenous surfactants. *Am Rev Respir Dis.* **145**: 24–30.

Hall, S. B., A. R. Venkitaraman et al. 1992. Importance of hydrophobic apoproteins as constituents of clinical exogenous surfactants. *Am Rev Respir Dis* **145**(1): 24–30.

Halpern, D. and J. B. Grotberg 1992. Fluid-elastic instabilities of liquid-lined flexible tubes. *J Fluid Mech* **244**: 615–632.

Halpern, D. and J. B. Grotberg 1993. Surfactant effects on fluid-elastic instabilities of liquid-lined flexible tubes: A model of airway closure. *J Biomech Eng* **115**: 271–277.

Hawco, M. W., K. P. Coolbear et al. 1981. Exclusion of fluid lipid during compression of monolayers of mixtures of dipalmitoylphosphatidylcholine with some other phosphatidylcholines. *Biochim Biophys Acta* **646**: 185–187.

Heil, M. 1999. Minimal liquid bridges in non-axisymmetrically buckled elastic tubes. *J Fluid Mech* **380**: 309–337.

Heil, M., A. L. Hazel et al. 2008. The mechanics of airway closure. *Respir Physiol Neurobiol* **163**(1–3): 214–221.

Heron, M., D. Hoyert et al. 2009. Deaths: Final data for 2006. *National Vital Statistics Reports*. Hyattsville, MD, National Center for Health Statistics.

Hill, M. J., T. A. Wilson et al. 1997. Effects of surface tension and intraluminal fluid on mechanics of small airways. *J Appl Physiol* **82**(1): 233–239.

Hughes, J. M., D. Y. Rosenzweig et al. 1970. Site of airway closure in excised dog lungs: Histologic demonstration. *J Appl Physiol* **29**(3): 340–344.

Jacob, A. M. and D. P. Gaver 2005. An investigation of the influence of cell topography on epithelial mechanical stresses during pulmonary airway reopening. *Phys Fluids* **17**(3): 031502 (11 pages).

Kamm, R. D. and R. C. Schroter 1989. Is airway closure caused by a liquid film instability? *Respir Physiol* **75**: 141–156.

Kay, S. S., A. M. Bilek et al. 2004. Pressure gradient, not exposure duration, determines the extent of epithelial cell damage in a model of pulmonary airway reopening. *J Appl Physiol* **97**: 269–276.

Kim, K. J. and A. B. Malik 2003. Protein transport across the lung epithelial barrier. *Am J Physiol Lung Cell Mol Physiol* **284**(2): L247–L259.

Klaus, M. A., J. A. Clements et al. 1961. Composition of surface-active material isolated from beef lung. *Proc Natl Acad Sci USA* **47**: 1858–1859.

Krueger, M. and D. Gaver 2000. A theoretical model of pulmonary surfactant multilayer collapse under oscillating area conditions. *J Colloid Interface Sci* **229**: 353–364.

Levitzky, M. 1982. *Pulmonary Physiology*. New York, McGraw Hill.

Macklem, P. T., D. F. Proctor et al. 1970. The stability of peripheral airways. *Respir Physiol* 8: 191–203.

Montgomery, A. B., M. A. Stager et al. 1985. Causes of mortality in patients with the adult respiratory-distress syndrome. *Am Rev Respir Dis* 132(3): 485–489.

Neergaard, K. V. 1929. Neue Auffassungen uber einen Grundbegriff der Atemmechanik. Die Retraktionskraft der Lunge, abhaenging von der Oberflaechenspannung in den Alveolen. *Z. Gesant Exp Med* 66: 373–394.

Otis, D. R., E. P. Ingenito et al. 1994. Dynamic surface-tension of surfactant Ta – experiments and theory. *J Appl Physiol* 77(6): 2681–2688.

Pillert, J. E. and D. P. Gaver 2009. Physicochemical effects enhance surfactant transport in pulsatile motion of a semi-infinite bubble. *Biophys J* 96(1): 312–327.

Reinelt, D. A. and P. G. Saffman 1985. The penetration of a finger into a viscous-fluid in a channel and tube. *SIAM J Sci Stat Comput* 6(3): 542–561.

Robertson, B. 1984. Lung surfactant. *Pulmonary Surfactant*. B. Robertson, L. V. Goulde and J. Batenburg. Amsterdam, Elsevier.

Rosenzweig, J. and O. E. Jensen 2002. Capillary-elastic instabilities of liquid-lined lung airways. *J Biomech Eng Trans ASME* 124(6): 650–655.

Ross, M. H. and W. Pawlina 2006. *Histology: A Text and Atlas: With Correlated Cell and Molecular Biology*. Baltimore, MD, Lippincott Williams & Wilkins.

Rubenfeld, G. D., E. Caldwell et al. 2005. Incidence and outcomes of acute lung injury. *New Engl J Med* 353(16): 1685–1693.

Schurch, S., F. H. Y. Green et al. 1998. Formation and structure of surface films: Captive bubble surfactometry. *Biochim Biophys Acta Mol Basis Dis* 1408(2–3): 180–202.

Smith, B. J. and D. P. Gaver 2008. The pulsatile propagation of a finger of air within a fluid-occluded cylindrical tube. *J Fluid Mech* 601: 1–23.

Syrkina, O., B. Jafari et al. 2008. Oxidant stress mediates inflammation and apoptosis in ventilator-induced lung injury. *Respirology* 13: 333–340.

Taylor, G. I. 1961. Deposition of a viscous fluid on the wall of a tube. *J Fluid Mech* 10(02): 161–165.

Tchoreloff, P., A. Gulik et al. 1991. A structural study of interfacial phospholipid and lung surfactant layers by transmission electron microscopy after Blodgett sampling: Influence of surface pressure and temperature. *Chem Phys Lipids* 59: 151–165.

Usher, R. H., A. C. Allen et al. 1971. Risk of respiratory distress syndrome related to gestational age, route of delivery, and maternal diabetes. *Am J Obstet Gynecol* 111: 826–832.

Vlahakis, N. E. and R. D. Hubmayr 2000. Cellular responses to mechanical stress invited review: Plasma membrane stress failure in alveolar epithelial cells. *J Appl Physiol* 89: 2490–2496.

Ware, L. B. and M. A. Matthay 2000. The acute respiratory distress syndrome. *N Engl J Med* 342: 1334–1349.

Yalcin, H. C., S. F. Perry et al. 2007. Influence of airway diameter and cell confluence on epithelial cell injury in an *in vitro* model of airway reopening. *J Appl Physiol* 103(5): 1796–1807.

Yu, S. and F. Possmayer 1993. Adsorption, compression and stability of surface films from natural, lipid extract and reconstituted pulmonary surfactants. *Biochim Biophys Acta* 1167: 264–271.

Zimmer, M. E., H. A. R. Williams et al. 2005. The pulsatile motion of a semi-infinite bubble in a channel: Flow fields, and transport of an inactive surface-associated contaminant. *J Fluid Mech* 537: 1–33.

Index

A

Active transport, 1-9
  biomimetics, 1-9
  SLMS, 1-9 to 1-10
Acute lung injury (ALI), 10-5 to 10-6
Acute or adult respiratory distress syndrome
    (ARDS), 10-5 to 10-6
Adenosine diphosphate (ADP), 6-5
Adenosine triphosphate (ATP), 6-1, 6-2
  Damkhöler number, 6-5
  transport characteristics in straight aorta, 6-7
Adherens junctions (AJs), 8-4, 8-5
ADMET (Adsorption–distribution–metabolism–
    elimination–toxicity), 5-8
ADP, see Adenosine diphosphate (ADP)
Adsorption–distribution–metabolism–
    elimination–toxicity, see ADMET
    (Adsorption–distribution–metabolism–
    elimination–toxicity)
Adsorptive-mediated pathway (AMT), 8-17 to 8-18
Adult respiratory distress syndrome, see Acute or adult
    respiratory distress syndrome (ARDS)
Airway closure, 10-6 to 10-7
Airway reopening, 10-7
  atelectatic airways reopening, 10-9
  capillary number, 10-8
  parallel-plate model of, 10-8
  shear stress, 10-9
  Stokes and continuity equations, 10-7
  stress exposure duration, 10-9
  stress gradient, 10-9
  stress tensor, 10-8
AJs, see Adherens junctions (AJs)
ALI, see Acute lung injury (ALI)
Alveolar-capillary barrier, 10-1
AMT, see Adsorptive-mediated pathway (AMT)
Animal surrogate systems, 5-1, 5-8; see also
    Cell culture analog (CCA)
  alternatives, 5-2
  barrier tissue models and μCCAs, 5-5 to 5-8
  cell culture analog concept, 5-2

  future prospects, 5-8
  limitations, 5-1
  prototype CCAs, 5-3 to 5-5
ARDS, see Acute or adult respiratory distress
    syndrome (ARDS)
Arterial wall mass transport, 6-1, 6-11 to 6-12; see also
    Atherogenesis
  blood phase transport role, 6-12 to 6-13
  Damkhöler numbers, 6-5 to 6-6
  nonuniform geometries, 6-7 to 6-11
  Sherwood numbers, 6-6 to 6-7
  solute concentration profile, 6-2
  steady-state transport modeling, 6-2 to 6-5
  transport characteristics in straight aorta, 6-7
Atherogenesis; see also Arterial wall mass transport
  bifurcation, 6-9 to 6-10
  blood phase transport role, 6-12 to 6-13
  curvature, 6-10 to 6-11
  flow separation zone, 6-9
  nonuniform geometries associated with, 6-7
  separation zone, 6-8
  stenosis, 6-8 to 6-9
  sudden expansion, 6-7 to 6-8
Atherosclerosis, 6-1
  fluid mechanics in 6-1
  lipid accumulation, 6-12
ATP, see Adenosine triphosphate (ATP)
Autoimmune disorder treatment, see Hormone disease
    control via tissue therapy

B

BAECs, see Bovine aortic endothelial cells (BAECs)
B[a]P, see Benzo[a]pyrene (B[a]P)
Basal metabolic rates, 2-2, 2-3
Basic fibroblast growth factor (bFGF), 1-8
BBB, see Blood–brain barrier (BBB)
BCEC, see Brain capillary endothelial cells (BCEC)
BCNU, see 1,3-bis(2-chloroethyl)-1-nitrosourea
    (BCNU)
BEI, see Brain efflux index (BEI)
Benzo[a]pyrene (B[a]P), 5-7

bFGF, *see* Basic fibroblast growth factor (bFGF)
Biofluid mechanics, 4-2, 7-9; *see also* Blood;
      Fluid dynamics
Biological processes, 7-9
Biomedical engineering (BME), 1-11
Biomedical reaction engineering, 7-1
Biomimetic membranes, 1-8
    active transport biomimetics, 1-9 to 1-10
    carrier protein, 1-8 to 1-9
    facilitated diffusion, 1-8, 1-10
    group translocation, 1-9
    for ion transport, 1-8 to 1-9
    jumping mechanism, 1-10 to 1-11
    lipid bilayer, 1-8
    passive diffusion, 1-8
    synthetic membranes, 1-9
Biomimetics, 7-2
    concepts, 1-2
    electro-enzymatic membrane reactors, 1-13
    mass transfer resistances assessment, 1-11
    membranes, 1-8
    and tissue engineering, 1-4
Biomimicry and tissue engineering, 1-4
    blood–brain barrier, 1-6 to 1-7
    implants, 1-7 to 1-8
    integrated systems, 1-5 to 1-6
    tissue reconstruction, 1-4
    vascular system, 1-7
Biomimicry concepts, 1-2
    bioreaction engineering, 1-4
    biotechnology, 1-4
    morphology and properties development, 1-3
    systems development, 1-4
    thin films and nanocapsule, 1-3 to 1-4
Biopolymers, 1-3
Bioreactors, 7-9; *see also* Reacting systems;
      Tissue microenvironments
    microreactor design, 7-10 to 7-11
    reactor types, 7-10
    scale-up and operational maps, 7-11
1,3-bis(2-chloroethyl)-1-nitrosourea (BCNU), 9-2
    diffuse through brain tissue, 9-10
    diffusion–elimination model, 9-7 to 9-9
    release, 9-2, 9-3
Blood; *see also* Biofluid mechanics;
      Tissue microenvironments
    circulation, 4-1
    flow, 4-2, 8-1
    fluid dynamics, 4-2
    hemoglobin binding characteristics, 7-3 to 7-4
    oxygen solubility, 7-3
    physiological respiratory function
      interpretation, 7-3
Blood phase transport, 6-12; *see also* Arterial wall
      mass transport; Atherogenesis
    direct mechanical effects, 6-12
    hypoxic effect, 6-12

    species flux, 6-2
    VEGF induction, 6-12 to 6-13
Blood–brain barrier (BBB), 1-2, 1-6, 8-2, 8-3, 9-1;
      see also Drug delivery; Brain
    anatomical structure 8-4
    characteristics of, 5-7
    coculture system, 8-11 to 8-12
    drug delivery, 1-6 to 1-7
    efflux systems, 8-6
    endothelial cells, 8-5
    glycocalyx layer, 8-3
    in human body, 8-11
    in vitro BBB model, 5-7
    in vitro models, 8-11 to 8-12
    junctional complexes, 8-4, 8-5
    pericytes, 8-3, 8-4
    and peripheral microvessels, 8-4
    permeability, 8-5, 8-7 to 8-10
    role, 8-3
    surface area of, 8-3
    transport models, 8-12 to 8-14
    transport pathways across, 8-5 to 8-6
    transwell system, 8-11
Blood–CSF barrier, 8-2; *see also* Blood–brain
      barrier (BBB)
BME, *see* Biomedical engineering (BME)
Bone marrow microenvironment, 7-4
Bovine aortic endothelial cells (BAECs), 6-11
Bovine serum albumin (BSA), 7-16
Brain, 8-1; *see also* Blood–brain barrier
      (BBB); Cerebrospinal fluid (CSF);
      Diffusion–elimination model; Interstitial
      transport
    allometry, 2-3
    barrier locations in, 8-2
    blood supply in cerebral cortex, 8-18
    capillaries permeability, 9-1 to 9-2
    diseases, 8-13
    drug delivery to, 8-19, 9-1
    drug uptake, 9-1
    endothelium, 8-6
    glial cells, 8-1
    infusion systems, 9-2
    neuroactive solutes, 8-3
    oxygen consumption, 2-3
    pial microvessels, 8-2
    polymer implant, 9-4
    total mass, 2-2
    vascular barrier system, 8-2
Brain capillary endothelial cells (BCEC), 1-7
    disadvantages, 8-12
    primary, 8-12
    for tissue culture systems, 8-11
Brain efflux index (BEI), 8-8
Brain uptake index (BUI), 8-8
Brain–CSF barrier, 8-2, 8-6; *see also* Brain
    circulation, 8-7

location, **8**-3
   total exchange area, **8**-7, **8**-18
Bronchioles, **10**-1; *see also* Lungs
BSA, *see* Bovine serum albumin (BSA)
BUI, *see* Brain uptake index (BUI)

C

Caco-2 cell model, **5**-6
Capillary number, **10**-8
Carrier protein, **1**-8 to **1**-9
CCA, *see* Cell culture analog (CCA)
CED, *see* Convection-enhanced distribution (CED)
Cell culture analog (CCA), **1**-2, **5**-1, **5**-2; *see also* Animal
         surrogate systems; Physiologically based
         pharmacokinetic models (PBPK models)
   barrier tissue models and use with, **5**-5 to **5**-8
   microfabricated devices, **5**-4
   microscale, **5**-4 to **5**-5
   μCCAs, **5**-5
   and PBPK, **5**-3
   prototype, **5**-3
Cell microenvironment manipulation, **7**-1; *see also*
         Tissue microenvironments
Cellular communication, **7**-4
Central nervous system (CNS), **8**-3; *see also* Brain
   barriers in, **8**-2
   BBB, **8**-14
   metastasis, **8**-19
Cerebral microvessels, **8**-1 to **8**-2, **8**-9; *see also* Brain;
         Pial microvessels
Cerebrospinal fluid (CSF), **8**-2, **8**-3, **8**-18;
         see also Brain
   brain–CSF barrier, **8**-2, **8**-6
   circulation, **8**-7
   drug delivery through, **8**-18 to **8**-20
   pharmacokinetic model for RIT delivered
         through, **8**-20
   production and circulation of, **8**-18
   radioimmunotherapy delivered through, **8**-18
   secretion, **8**-6
   tumor cells, **8**-19
CFD, *see* Computational fluid dynamics (CFD)
Chained carrier, **1**-10
Charged cell-penetrating peptides (CPPs), **8**-17 to **8**-18
Chemical reactions, **7**-9
Choroid epithelium, **8**-18
Choroid plexus, **8**-2
Circulatory system, **4**-6
CNS, *see* Central nervous system (CNS)
Coenzyme, **1**-13
Collagen type IV network, **8**-12
Colony-stimulating factors, **7**-4
Computational fluid dynamics (CFD), **4**-3
   applications, **4**-15 to **4**-16
   for *in vivo* implantation, **7**-9
Conduction shape factor, **3**-13

Connexon, **7**-5
Conservation equations, **4**-4
   conservation of mass, **4**-4 to **4**-5
   continuity condition, **4**-4
Continuous stirred tank reactors (CSTRs), **2**-7
Continuum mechanics elements, **4**-3; *see also*
         Hydrodynamics elements, theoretical
   conservation equations, **4**-4 to **4**-5
   constitutive equations, **4**-3
   Newtonian, **4**-3
   turbulence and instabilities, **4**-5 to **4**-6
Continuum models, **3**-6; *see also* Perfused tissue models
   combination, **3**-8
   directed perfusion, **3**-7
   effective conductivity model, **3**-7 to **3**-8
   formulations, **3**-6
   heat sink model, **3**-6 to **3**-9
Controlled release system, **9**-2, **9**-12; *see also*
         Drug delivery
   anticancer compounds, **9**-3
   conservation equation, **9**-2 to **9**-3
   diffusion-mediated release, **9**-2
Convection-enhanced distribution (CED), **8**-14
Convective–diffusive mechanism, **6**-2
Countercurrent heat exchange, **3**-5; *see also*
         Vascular models
CPPs, *see* Charged cell-penetrating peptides (CPPs)
CSF, *see* Cerebrospinal fluid (CSF)
CSTRs, *see* Continuous stirred tank reactors (CSTRs)
Cytovene, **8**-16

D

Damkhöler number, **6**-3; *see also* Arterial wall
         mass transport
   adenosine triphosphate, **6**-5
   albumin and LDL, **6**-5 to **6**-6
   oxygen, **6**-6
   for solutes, **6**-5
Diabetes, **7**-13
Dialysis model, **2**-11; *see also* Hemodialysis
Diffusion-controlled reaction, **2**-12; *see also*
         Gene expression
Diffusion–elimination model, **9**-9; *see also*
         Drug delivery
   application, **9**-7 to **9**-9
   drug release rate effect, **9**-9
   failure situations, **9**-9
   fluid convection effect, **9**-10
   interstitial fluid velocity, **9**-11
   limitations and extensions of, **9**-9
   metabolism effect, **9**-11 to **9**-12
   tissue penetration determinants, **9**-9 to **9**-10
Diffusive permeability, **8**-8
Dipalmitoyl phosphatidylcholine (DPPC), **10**-4
Direct cell-to-cell contact **7**-5
Dispersion models, **2**-13

DPPC, *see* Dipalmitoyl phosphatidylcholine (DPPC)
Drug delivery, 8-7, 8-14; *see also* Blood–brain
barrier (BBB); Cerebrospinal fluid (CSF);
Diffusion–elimination model
adsorptive-mediated pathway, 8-17 to 8-18
through blood–brain barrier, 8-7, 8-14
brain, 9-1 to 9-2
CED methods, 8-14 to 8-15
through cerebrospinal fluid, 8-18 to 8-20
controlled release system, 9-2 to 9-3
diffusion coefficients, 9-3
drug elimination rate, 9-5
drug transport after release from implant,
9-4 to 9-7
implant design, 9-8
infusion systems, 9-2
interstitial fluid velocity, 9-11
intracranial BCNU delivery systems, 9-7 to 9-9
lipophilic diffusion pathway, 8-16
microinfusion, 9-10
new approaches to, 9-12
paracellular pathway, 8-16
receptor-mediated pathway, 8-17
transport and, 8-1
transporter-mediated pathway, 8-16 to 8-17

E

ECF, *see* Extracellular fluid (ECF)
ECM, *see* Extracellular matrix (ECM)
ECS, *see* Extracellular space (ECS)
ECVAM, *see* European Center for the Validation of
Alternative Methods (ECVAM)
Effective conductivity, 3-13
model, 3-7
thermally equivalent fiber, 3-8
Electro-enzymatic membrane reactors, 1-13; *see also*
Biomimetics
lactate production, 1-13
in vivo coenzyme regeneration processes
mimicry, 1-13
Encapsulation motif, 7-13
diffusivity, 7-15
immunoisolative capabilities of, 7-13
membrane permeability, 7-15
physical and transport parameters, 7-15 to 7-16
shed antigens, 7-14
transport, 7-14
Endothelial cells, 7-7
Engineered membrane mimetics, 1-2
Engineered tissues, 5-8
Enzyme-catalyzed surface reaction, 6-2
EpC, *see* Epithelial cells (EpC)
EpiSkin model, 5-6
Epithelial cells (EpC), 10-5
damage and stress, 10-10
lining, 10-8

Equilibration lengths, 3-4, 3-13; *see also*
Vascular models
European Center for the Validation of Alternative
Methods (ECVAM), 5-6
Extracellular fluid (ECF), 9-4
Extracellular matrix (ECM), 1-3
and cell–tissue interactions, 7-5
Extracellular space (ECS), 9-4

F

Facilitated diffusion, 1-8, 1-10
FITC, *see* Fluorescein isothiocyanate (FITC)
Flow fields, 4-5 to 4-6
Flow in tubes, 4-6; *see also* Hydrodynamics
elements, theoretical
entrance flow, 4-7 to 4-8
fluid system energy equation, 4-8
head loss, 4-8
mechanical energy equation, 4-8 to 4-9
Navier–Stokes equations, 4-6
parabolic velocity profile characteristic, 4-7
pipe friction factor, 4-8 to 4-9
Poiseuille flow, 4-6 to 4-7
Flow reactor systems, 7-10; *see also* Bioreactors
Flow separation zone, 6-9
Fluid dynamics, 4-1; *see also* Biofluid mechanics;
Hydrodynamics elements, theoretical;
Navier–Stokes equations; Pulsatile flow
applications, 4-2 to 4-3
approximations to Navier–Stokes equations,
4-14 to 4-15
in atherosclerosis, 6-1
blood, 4-2
computational fluid dynamics, 4-15
fluid system energy equation, 4-8
fluid travel pathways, 4-2
models and computational techniques, 4-14
pulsatile flow, 4-9 to 4-14
synovial fluid, 4-2
theoretical hydrodynamics elements,
4-3 to 4-9
Fluid–structure interactions, 10-6; *see also*
Liquid lining flows; Lungs
airway closure, 10-6 to 10-7
airway reopening, 10-7 to 10-9, 10-10
biological responses to micromechanical
stress field, 10-9 to 10-11
stress boundary condition, 10-12
surfactant physicochemical interactions,
10-11 to 10-13
Fluorescein isothiocyanate (FITC), 8-12

G

γ-glutamyl-transpeptidase (γ-GTP), 8-11 to 8-12
γ-GTP, see γ-glutamyl-transpeptidase (γ-GTP)

Gene expression, **2**-11
    in bacterial cell, **2**-12
    diffusion-controlled reaction, **2**-12
    dispersion models, **2**-13
GFAP, *see* Glial fibrillary acidic protein (GFAP)
Glial cells, **8**-1
Glial fibrillary acidic protein (GFAP), **8**-11
Glucose dynamics, **2**-14
Glucose tolerance test, **2**-14
Glycocalyx layer, **8**-3
Glycolysis, **2**-9
Graetz number, **3**-6
Group translocation, **1**-9
Growth factors **7**-4 to **7**-5

H

HA, *see* Hyaluronic acid (HA)
Head loss, **4**-8
Heat transport, **3**-2
Hemodialysis, **2**-10, **2**-11
Hormone disease control via tissue therapy, **7**-11;
        see also Tissue microenvironments
    case study, **7**-13
    encapsulation motif, **7**-13 to **7**-16
    key issues, **7**-11
    limitations, **7**-12
    transport considerations, **7**-12 to **7**-13
Human body, **4**-1
Human vasculature endothelial cells
        (HUVECs), **5**-5
HUVECs, *see* Human vasculature endothelial
        cells (HUVECs)
Hyaluronic acid (HA), **1**-8
Hydrodynamic phenomena, **4**-1
Hydrodynamics elements, theoretical, **4**-3; *see also*
        Fluid dynamics
    continuum mechanics elements, **4**-3 to **4**-6
    flow in tubes, **4**-6
Hypoxia, **6**-12

I

Imaging techniques, **8**-8
Infant respiratory distress syndrome (IRDS), **10**-5
Inflammatory agents, **8**-16
Infusion systems, **9**-2
Integrated devices, **5**-8
Interstitial transport, **9**-1; *see also* Brain; Drug delivery
    diffusion–elimination models, **9**-7
    drug transport, **9**-4
    implantable controlled delivery systems, **9**-2
Intracerebral microdialysis, **8**-8
Intraventricular therapy, **9**-2
Inulin, **1**-7
Ion penetration time, **1**-11
IRDS, *see* Infant respiratory distress syndrome (IRDS)

J

Jumping mechanism, **1**-10
    ion penetration time, **1**-11
Junctional complexes, **8**-4, **8**-5

K

Kedem–Katchalsky equations, **8**-7

L

Lactate dihydrogenase (LDH), **5**-3
Lactate production, **1**-13
Laminar flow, **4**-5
Langmuir–Blodgett (LB), **1**-3
LB, *see* Langmuir–Blodgett (LB)
LDH, *see* Lactate dihydrogenase (LDH)
L-3,4-dihydroxyphenylalanine, *see* L-Dopa
        (L-3,4-dihydroxyphenylalanine)
LDL, *see* Low-density lipoprotein (LDL)
L-Dopa (L-3,4-dihydroxyphenylalanine), **8**-16, **9**-1
Leptomeninges (LM), **8**-19
LipDH, *see* Lipoamide dehydrogenase (LipDH)
Lipid bilayer, **1**-8; *see also* Biomimetic membranes
    synthetic, **1**-9
Lipoamide dehydrogenase (LipDH), **1**-13
Lipophilic diffusion pathway, **8**-16
Liquid lining flows, **10**-5; *see also* Fluid–structure
        interactions
    ALI–ARDS, **10**-5 to **10**-6
    IRDS, **10**-5
    VILI, **10**-6
LM, *see* Leptomeninges (LM)
Low-density lipoprotein (LDL), **6**-1, **8**-6
Lungs, **2**-16, **10**-1; *see also* Fluid–structure interactions;
        Liquid lining flows; Pulmonary
    alveolar-capillary barrier, **10**-1
    bifurcating airways, **10**-1
    pressure–volume hysteresis loops, **10**-2
    primary function, **10**-1
    pulmonary surfactant, **10**-3
    respiratory illnesses, **10**-5
    static pressure difference, **10**-2
    surface tension forces, **10**-2

M

mAbs, *see* Monoclonal antibodies (mAbs)
Macromolecules, **1**-8
Macroscopic approximations, **2**-2; *see also* Transport/
        reaction processes
    basal metabolic rates, **2**-2, **2**-3
    brain allometry, **2**-3
    total brain mass, **2**-2
Madin–Darby canine kidney, *see* MDCK (Madin–Darby
        canine kidney)

Magnetic resonance imaging (MRI), **8**-8
Major histocompatibility complex (MHC), 7-14
Mammalian topology, **2**-2
Marker species, 7-16
Mass transfer, 7-8
Mass transfer resistances assessment, **1**-11; *see also*
       Biomimetics
   in cell culture analog systems, 1-12 to 1-13
   in PBPK models, 1-12 to 1-13
   uncoupling resistances, 1-11 to 1-12
Mass transport, 3-1
3-MC, *see* 3-Methylcholanthrene (3-MC)
MDCK (Madin–Darby canine kidney), **5**-8
Mean residence time, 2-4
Membrane permeability, 7-15
Mesenchymal cells, 7-7
Metabolic process control, 7-1
Methanogenic chemo-autotrophic process, 1-12
Methotrexate (MTX), **9**-12
3-Methylcholanthrene (3-MC), 5-7
MHC, *see* Major histocompatibility complex (MHC)
Microcirculation, 2-6
Microencapsulation motif, 1-5, 7-2
Microphysiometer, 5-2
Microreactor, 7-10 to 7-11
Microvascular heat transfer, 3-1
   challenges, 3-1 to **3**-2
   clinical heat generation, 3-11
   concepts, 3-2 to 3-3
   model solutions, 3-11 to 3-13
   models for, 7-9
   parameter values, 3-9
   perfused tissue models, 3-6 to 3-9
   thermal properties, 3-9
   thermoregulation, 3-10 to 3-11
   vascular models, 3-3 to 3-6
Microvessel endothelium, **8**-7
Moen–Korteweg formula, 4-12
Molecular dynamics, 7-3
Molecular engineering, 7-1
Molecular Trojan horse, **8**-17
Monad models, **1**-12
Monoclonal antibodies (mAbs), **8**-19
MRI, *see* Magnetic resonance imaging (MRI)
MRP, *see* Multidrug resistance-associated
       proteins (MRP)
MTX, *see* Methotrexate (MTX)
MTX–dextran conjugates, **9**-12
Multidrug resistance-associated proteins (MRP), **8**-6
Multiequation models, 3-9; *see also* Perfused
       tissue models
Multiple time constant systems, 2-7; *see also*
       Transport/reaction processes
   boundary conditions, 2-7 to 2-8
   exercise and type II diabetes, 2-14
   effect of exercise vs. metformin therapy, 2-14 to 2-15
   gene expression in prokaryotes, 2-11 to 2-13
   glucose dynamics, 2-14
   glucose tolerance test, 2-14
   glycolysis, 2-9
   hemodialysis, 2-10 to 2-11
   pharmacokinetics, 2-8 to 2-10
   reactions controlling, 2-8
Murray's law, 2-5

N

Nanotechnologies, 1-3
National Institute of Diabetes and Digestive and
       Kidney Diseases (NIDDKD), 7-13
Navier–Stokes equations, 4-3; *see also* Fluid dynamics
   approximations to, 4-14 to 4-15
   in Cartesian coordinates, 4-4
   flow in pipe, 4-6
   for incompressible fluid, 4-4
   solution of, 4-4 to 4-5
Neovascularization index (NI), **1**-8
Neuroactive solutes, **8**-3
Neuroblastoma, **8**-19
Newtonian, 4-3
Newton's second law, 4-3
NI, *see* Neovascularization index (NI)
NIDDKD, *see* National Institute of Diabetes and
       Digestive and Kidney Diseases (NIDDKD)
Nonpolar molecules, 1-9
Normal-stress gradient, **10**-9
No-slip condition, 4-5
Nusselt number, 3-5

O

Ommaya reservoir, **8**-18
Operational maps, 7-11
Orders of magnitude, 2-3
Oxygen
   barrier, **6**-6
   carriers, 7-14
   Damkhöler number, **6**-6
   solubility, 7-3
   transport, **6**-4

P

Paracellular pathway, **8**-16
Passive diffusion, **1**-8
PBPK models, *see* Physiologically based
       pharmacokinetic models (PBPK models)
Peclet number, **10**-11
Pennes' bioheat equation, 3-6
   one-dimensional steady-state solution of, 3-12
   transient solution of, 3-12 to 3-13
Pennes' heat sink model, 3-6
   advantages of, 3-6
   limitation, 3-7

numerical solutions of, **3**-11
　　Pennes' bioheat equation, **3**-6
Perflourocarbons (PFC), **1**-3
Perfused tissue models, **3**-6; *see also*
　　　　Microvascular heat transfer
　　continuum models, **3**-6 to **3**-9
　　multiequation models, **3**-9
　　vascular reconstruction models, **3**-9
Perfusion, **4**-2
　　rate, **3**-13
　　systems, **4**-2
Pericytes, **8**-3, **8**-4
Peripheral microvessel, **8**-3
　　anatomical structure **8**-4
　　and BBB, **8**-4
　　BBB endothelial cells and, **8**-5
　　solute exchange, **8**-1
Permeability indicators, **8**-8
PET, *see* Positron emission tomography (PET)
PFC, *see* Perflourocarbons (PFC)
PFRs, *see* Plug flow reactors (PFRs)
P-glycoprotein (P-gp), **8**-6
P-gp, *see* P-glycoprotein (P-gp)
Pharmaceutical engineering, **7**-1
Pharmacokinetics, **2**-8
　　approximations, **2**-9
　　methods, **8**-8
　　model, **8**-20
　　PBPK, **5**-1, **5**-8
　　for RIT delivered through, **8**-20
Physiologically based pharmacokinetic models
　　　　(PBPK models), **1**-12, **5**-1, **5**-8; *see also*
　　　　Cell culture analog (CCA)
PI, *see* Process intensification (PI)
Pial microvessels, **8**-2; *see also* Brain
　　in vivo permeability measurement of, **8**-9
　　permeability, **8**-10
　　quantitative fluorescence imaging
　　　　method, **8**-10
Pipe friction factor, **4**-8 to **4**-9
Plug flow reactors (PFRs), **2**-7
Pluripotential stem cells, **7**-7
PMF, *see* Proton motive force (PMF)
Poiseuille flow, **4**-6, **4**-7
Poiseuille velocity profile, **4**-7
Poiseuille's law, **4**-6
Polar molecules, **1**-9
Polycationic proteins, **8**-17
Polymer blends, **1**-3
Positron emission tomography (PET), **8**-8
Process intensification (PI), **7**-11
Proton motive force (PMF), **1**-4
Pseudocontinuum models, **2**-15; *see also*
　　　　Transport/reaction processes
　　pulmonary blood-gas matching, **2**-17
　　pulmonary structure and function, **2**-16
　　tissue oxygenation, **2**-15

Pulmonary
　　blood vessels, **2**-16
　　blood-gas matching, **2**-17
　　net transport, **2**-16
　　organization oxygenation, **2**-17
　　structure and function, **2**-16
　　surfactant, **10**-3
Pulsatile flow, **4**-9; *see also* Fluid dynamics
　　hemodynamics, **4**-9 to **4**-11, **4**-11 to **4**-14
　　local bulging of tube wall, **4**-12
　　oscillatory components, **4**-10
　　oscillatory shear stress, **4**-11
　　representative velocity profiles, **4**-10 to **4**-11
　　steady components, **4**-10
　　turbulence in, **4**-14
Pyruvate dehydrogenase complex, **2**-7

R

Radioimmunotherapy (RIT), **8**-19
Reacting systems, **7**-9; *see also* Bioreactors;
　　　　Tissue microenvironments
Reactive species, highly, **6**-11
Real time polymerase chain reaction (RT-PCR), **8**-12
Receptor-mediated transcytosis (RMT), **8**-6, **8**-17
Regeneration system, **1**-13
Respiratory illnesses **10**-5
Reynolds number, **4**-5, **4**-14
RIT, *see* Radioimmunotherapy (RIT)
RMT, *see* Receptor-mediated transcytosis (RMT)
RNAi, *see* RNA interference (RNAi)
RNA interference (RNAi), **8**-14
RT-PCR, *see* Real time polymerase chain
　　　　reaction (RT-PCR)

S

Satellite cells, **7**-7
Separation zone, **6**-8
SGL, *see* Surface glycocalyx layer (SGL)
Shear stress, **10**-9, **10**-10
Sherwood number, **6**-3; *see also* Arterial wall
　　　　mass transport
　　in circulation, **6**-6
　　spatial distribution of, **6**-8, **6**-9, **6**-10
　　straight vessels, **6**-6 to **6**-7
SLMS, *see* Supported liquid membrane
　　　　systems (SLMS)
Specific heat, **3**-13
Steady Poiseuille flow, *see* Poiseuille flow
Steady-state transport modeling, **6**-2; *see also* Arterial
　　　　wall mass transport
　　Damköhler number, **6**-3
　　permeable surface, **6**-4
　　reactive surface, **6**-2 to **6**-3
　　reactive wall, **6**-4 to **6**-5
　　Sherwood number, **6**-3

Steady-state transport modeling (*Continued*)
  solute concentration profile, **6**-2
  transport to surface, **6**-3
  transport-limited, **6**-3
  wall-limited process, **6**-3
Stokes and continuity equations, **10**-7
Stokes layer, **4**-9
Stokes–Einstein equation, **7**-15
Stress exposure duration, **10**-9
Supported liquid membrane systems (SLMS),
    1-9 to **1**-10; *see also* Active transport
Surface glycocalyx layer (SGL), **8**-13
Surfactant, **10**-4
  flux, **10**-11
  monolayer, **10**-4
  physicochemical interactions, **10**-11 to **10**-13
  pulmonary, **10**-3
  transport and fluid mechanical properties, **10**-4;
      see also Fluid–structure interactions
Synovial fluid, **4**-2
Synthetic polymers, **1**-3, **1**-9
System repair agents, **4**-1

T

TER, *see* Transepithelial electrical
      resistance (TER)
TEER, *see* Transendothelial electrical
      resistance (TEER)
Thermal conductivity, **3**-13
Thermal therapies, **3**-1, **3**-11
Thermally significant blood vessels, **3**-2, **3**-4, **3**-13
Thermodynamics, first law of, **4**-8
Thermoregulation, **3**-10
  metabolic rate, **3**-11
  temperature dependence of blood perfusion
      effect, **3**-10
Tight junctions (TJs), **8**-4, **8**-5
Time constant, **2**-4; *see also* Transport/reaction
      processes
  alveolar transients, **2**-5
  diffusion lengths, **2**-4
  intracellular diffusion coefficients, **2**-4
  mean residence time, **2**-4
  types of, **2**-4
Time constant ratios, **2**-5; *see also* Transport/reaction
      processes
  blood vessel classification, **2**-5 to **2**-6
  effectiveness factors, **2**-6
  pyruvate dehydrogenase complex, **2**-7
  simultaneous diffusion and chemical reaction,
      2-6 to **2**-7
Tissue, **2**-16
  oxygenation, **2**-15
  reconstruction, **1**-4; *see also* Biomimicry and
      tissue engineering
  slice, **5**-8

Tissue microenvironments, **7**-1 to **7**-2, **7**-3; *see also*
      Bioreactors; Reacting systems
  cellular communication, **7**-4
  cellularity, **7**-6 to **7**-7
  communication, **7**-4 to **7**-6
  communication with whole-body
      environment, **7**-6
  direct cell-to-cell contact **7**-5
  dynamics, **7**-7
  ECM and cell–tissue interactions, **7**-5
  engineering, **7**-2
  geometry, **7**-7 to **7**-8
  growth factors **7**-4 to **7**-5
  microcirculation, **7**-2
  performance criteria specification, **7**-3
  system interactions, **7**-8 to **7**-9
  tissue function estimation, **7**-3 to **7**-4
Tissue thermal properties, **3**-9
  representative thermal property values, **3**-10
TJs, *see* Tight junctions (TJs)
TOA, *see* Tri-*n*-octylamine (TOA)
Transendothelial electrical resistance
      (TEER), **1**-7, **8**-11
Transepithelial electrical resistance (TER), **5**-8
Transport phenomena, **7**-2
Transport/reaction processes, **2**-1
  cellular crowding, **2**-18
  complex situations, **2**-17
  dual nature of oxygen, **2**-18
  genetic regulation stochastic behavior, **2**-17
  macroscopic approximations, **2**-2 to **2**-3
  mammalian topology, **2**-2
  multiple time constants, **2**-7 to **2**-15
  orders of magnitude, **2**-3
  pseudocontinuum models, **2**-15 to **2**-17
  self-organization and emergence, **2**-18
  time constant ratios, **2**-5 to **2**-7
  time constants, **2**-4 to **2**-5
Transporter-mediated pathway, **8**-16 to **8**-17
Transwell system, **8**-11
Tri-*n*-octylamine (TOA), **1**-10
Turbulent flows, **4**-5
  closure problem of turbulence, **4**-6

V

Vascular endothelial-cadherin (VE-cadherin), **8**-4
Vascular endothelial growth factor
      (VEGF), **1**-8, **6**-12
Vascular models, **3**-3; *see also* Microvascular
      heat transfer
  countercurrent heat exchange, **3**-5
  equilibration lengths, **3**-4
  heat transfer, **3**-5 to **3**-6
  representative tissue cylinder surrounding
      blood vessel, **3**-4
  vascular geometries and shape factors, **3**-5

Vascular reconstruction models, **3**-9; *see also*
    Perfused tissue models
Vascular system, **7**-6
Vascular tissues, **3**-2
    blood and tissue subvolumes, **3**-2
    heat transfer, **3**-3

VE-cadherin, *see* Vascular endothelial-cadherin
    (VE-cadherin)
VEGF, *see* Vascular endothelial growth
    factor (VEGF)
Ventilator-induced lung injury (VILI), **10**-6
VILI, *see* Ventilator-induced lung injury (VILI)